NEW FORMAT

AJ Sadler

Mathematics Applications

2

Student Book

Unit 2

Mathematics Applications Unit 2
1st Edition
A.J. Sadler

Publishing editor: Robert Yen
Project editor: Alan Stewart
Cover design: Chris Starr (MakeWork)
Text designers: Sarah Anderson, Nicole Melbourne,
Danielle Maccarone
Permissions researcher: Lyahna Spencer
Answer checker: George Dimitriadis
Production controller: Erin Dowling
Typeset by: Cenveo Publisher Services

Any URLs contained in this publication were checked for currency
during the production process. Note, however, that the publisher
cannot vouch for the ongoing currency of URLs.

© 2016 A.J. Sadler

For product information and technology assistance,
in Australia call **1300 790 853**;
in New Zealand call **0800 449 725**

For permission to use material from this text or product, please email
aust.permissions@cengage.com

National Library of Australia Cataloguing-in-Publication Data
Sadler, A.J., author.
Mathematics applications : unit 2 / A.J. Sadler.

1st revised edition
9780170390262 (paperback)
Includes index.
For secondary school age.

Mathematics--Study and teaching (Secondary)--Western Australia.
Mathematics--Textbooks.

510.712

Cengage Learning Australia
Level 7, 80 Dorcas Street
South Melbourne, Victoria Australia 3205

Cengage Learning New Zealand
Unit 4B Rosedale Office Park
331 Rosedale Road, Albany, North Shore 0632, NZ

For learning solutions, visit **cengage.com.au**

Printed in China by 1010 Printing International Limited.
12 13 14 15 26 25 24

PREFACE

This text targets Unit Two of the West Australian course *Mathematics Applications*, a course that is organised into four units altogether, the first two for year eleven and the last two for year twelve.

The West Australian course, *Mathematics Applications*, is based on the Australian Curriculum Senior Secondary course *General Mathematics*. The main difference between Unit Two of these two courses is the inclusion of some work on the Normal Distribution in the West Australian course, covered in the final chapter of this text. Hence, by excluding the final chapter this text is also suitable for anyone following Unit Two of the Australian Curriculum course, *General Mathematics*.

The book contains text, examples and exercises containing many carefully graded questions. A student who studies the appropriate text and relevant examples should make good progress with the exercise that follows.

The book commences with a section entitled **Preliminary work**. This section briefly outlines work of particular relevance to this unit that students should either already have some familiarity with from the mathematics studied in earlier years, or for which the brief outline included in the section may be sufficient to bring the understanding of the concept up to the necessary level.

As students progress through the book they will encounter questions involving this preliminary work in the **Miscellaneous Exercises** that feature at the end of each chapter. These miscellaneous exercises also include questions involving work from preceding chapters to encourage the continual revision needed throughout the unit.

Some chapters commence with a **'Situation'** or two for students to consider, either individually or as a group. In this way students are encouraged to think and discuss a situation, which they are able to tackle using their existing knowledge, but which acts as a forerunner and stimulus for the ideas that follow. Students should be encouraged to discuss their solutions and answers to these situations and perhaps to present their method of solution to others. For this reason answers to these situations are generally not included in the book.

At times in this series of books I have found it appropriate to go a little outside the confines of the syllabus for the unit involved. In this regard readers will find in this text I have included some consideration of pie charts as a method of data display. Similarly, with Linear Relationships, whilst the syllabus concentrates on determining the slope and intercept from the equation or plot, I also include tables of values and include determining the equation knowing the gradient and one other point on the line or knowing two points on the line. When using the sine rule to determine the size of an unknown angle in non right triangles, I do not confine consideration to acute angles. However, when an obtuse angle is involved, the reader is told this fact so that ambiguous situations are still avoided.

Alan Sadler

ISBN 9780170390262

CONTENTS

PRELIMINARY WORK · vii

Use of number vii

Ratios vii

Coordinates vii

Data display............................ viii

Data analysis........................... viii

Formulae x

Algebra x

Similar triangles........................ xi

The Pythagorean theorem.............. xi

Probability................................ xi

Geometry xi

Use of technology to process
and display data xii

1

UNIVARIATE DATA: CLASSIFY, ORGANISE AND DISPLAY · 2

Types of data............................. 4

Variables 4

Categorical variables................... 4

Displaying categorical data............ 6

Numerical variables..................... 8

Displaying numerical data.............. 9

Histograms and bar charts 11

Miscellaneous exercise one 17

2

SUMMARISING DATA AND DESCRIBING DISTRIBUTIONS · 20

Combining groups...................... 24

Use of statistical functions on
a calculator 30

Grouped data 31

Describing a distribution of scores... 36

Miscellaneous exercise two 39

3

MEASURES OF DISPERSION OR SPREAD · 42

Standard deviation 45

Use of statistical functions on
a calculator 47

Frequency tables 53

Outliers 53

Grouped data 54

Central tendency and spread
– An investigation....................... 57

Miscellaneous exercise three 58

4

BOXPLOTS, HISTOGRAMS AND MORE ABOUT DESCRIBING DISTRIBUTIONS · 62

Box and whisker diagrams
(boxplots) 63

Boxplot or histogram?.................. 67

More about the shape of a
distribution – skewness 68

Miscellaneous exercise four 74

5

THE STATISTICAL INVESTIGATION PROCESS · 78

Implementing the statistical
investigation process 80

Miscellaneous exercise five 81

6

SOLVING EQUATIONS · 84

Solving equations 86

Equations with brackets or
fractions................................... 88

Linear equations 92

Miscellaneous exercise six............ 96

7

USING EQUATIONS TO SOLVE PROBLEMS 100
Pyramids.................................. 101
Number puzzles........................ 105
Solving problems....................... 109
Equations from simple interest formula.................................... 113
Equations from ratios.................. 114
Miscellaneous exercise seven 121

8

LINEAR RELATIONSHIPS 124
Straight line graphs.................... 127
The gradient of a straight line 127
Table of values.......................... 128
What's my rule? 132
Table – rule – graph................... 133
Lines parallel to the axes 135
Use of a calculator with a graphing facility 136
More about $y = mx + c$, the equation of a straight line............ 142
It may not look like $y = mx + c$ but it may still be linear. 143
Determining the equation of a straight line......................... 143
A useful rule 144
Calculator routines.................... 144
Linear relationships in practical situations................................. 146
Miscellaneous exercise eight 152

9

PIECEWISE DEFINED RELATIONSHIPS 156
Piecewise defined relationships 158
Miscellaneous exercise nine 163

10

TRIGONOMETRY FOR RIGHT TRIANGLES 166
Right angled triangles................. 168
Trigonometry............................ 170
Hypotenuse, opposite and adjacent.................................. 171
Notes regarding calculator usage....................................... 174
Applications 176
Accuracy and trigonometry questions 183
Bearings 184
Elevation and depression 184
More vocabulary....................... 185
Miscellaneous exercise ten 188

11

TRIGONOMETRY FOR TRIANGLES THAT ARE NOT RIGHT ANGLED 192
Area of a triangle...................... 193
Triangles that are not right angled 194
Area of a triangle given two sides and the included angle 195
The sine rule............................. 200
The cosine rule......................... 204
Miscellaneous exercise eleven 213

12

SIMULTANEOUS LINEAR EQUATIONS 218
Introducing two variables 219
Solving word problems............... 225
Miscellaneous exercise twelve 231

13

STANDARD SCORES AND THE NORMAL DISTRIBUTION 236
Standard scores 238
Normal distribution 240
Using a calculator 242
In the old days: Using a book of tables 242
Notation.................................. 245
Quantiles................................. 247
Miscellaneous exercise thirteen 251

ANSWERS 257
INDEX.................................... 282

IMPORTANT NOTE

This series of texts has been written based on my interpretation of the appropriate *General Mathematics* syllabus documents (Australian Curriculum) and *Mathematics Applications* syllabus documents (Western Australia) as they stand at the time of writing.
It is likely that as time progresses some points of interpretation will become clarified and perhaps even some changes could be made to the original syllabus. I urge teachers of these courses, and students following the courses, to check with the appropriate curriculum authority to make themselves aware of the latest version of the syllabus current at the time they are studying the course.

Acknowledgements

As with all of my previous books I am again indebted to my wife, Rosemary, for her assistance, encouragement and help at every stage.

To my three beautiful daughters, Rosalyn, Jennifer and Donelle, thank you for the continued understanding you show when I am 'still doing sums' and for the love and belief you show in me.

Further acknowledgements

I wish to thank the following companies and organisations for allowing me to use some of their data as the basis for some of the questions, and occasionally in the theory sections, of this text:

Australian Bureau of Statistics

Bureau of Meteorology

National Heart Foundation of Australia

The New Internationalist Magazine

The Western Australian Police

West Australian Department of Education and Training

Alan Sadler

PRELIMINARY W●RK

This book assumes that you are already familiar with a number of mathematical ideas from your mathematical studies in earlier years.

This section outlines the ideas which are of particular relevance to Unit Two of the *Mathematics Applications* course and for which some familiarity will be assumed, or for which the brief explanation given here may be sufficient to bring your understanding of the concept up to the necessary level.

Read this 'Preliminary work' section and if anything is not familiar to you, and you don't understand the brief explanation given here, you may need to do some further reading to bring your understanding of those concepts up to an appropriate level for this unit. (If you do understand the work but feel somewhat 'rusty' with regards to applying the ideas some of the chapters afford further opportunities for revision, as do some of the questions in the miscellaneous exercises at the end of chapters.)

- Chapters in this book will continue some of the topics from this preliminary work by building on the assumed familiarity with the work.

- The miscellaneous exercises that feature at the end of each chapter may include questions requiring an understanding of the topics briefly explained here.

Use of number

The understanding and appropriate use of the rule of order, fractions, decimals, percentages, rounding, truncation, square roots and cube roots, numbers expressed with positive integer powers, e.g. 2^3, 5^2, 2^5 etc, expressing numbers in standard form, e.g. 2.3×10^4 $(= 23\,000)$, 5.43×10^{-7} $(= 0.000\,000\,543)$, also called scientific notation, and familiarity with the symbols $>$, $\geq$, $<$, and $\leq$ is assumed.

Ratios

The idea of comparing two or more quantities as a *ratio* should be something you are familiar with.

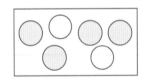

For example, for the diagram on the right the ratio of unshaded circles to shaded circles is $2 : 4$ which simplifies to $1 : 2$.

Suppose the ratio of males to females in a school is $17 : 21$.

If we know that there are 231 females in the school we can determine the number of males.

Males : females $= 17 : 21$
$ = \ ?: 231$ $\Big\}\times 11$

The number of males $= 17 \times 11$, i.e. 187.

```
231 / 21
                    11
```

Coordinates

It is assumed that you are familiar with the idea that points on a graph can be located by stating the coordinates of the point.

For example   point A   has coordinates $(3, 2)$,
              point B   has coordinates $(-2, 3)$,
              point C   has coordinates $(-3, -2)$,
              point D   has coordinates $(2, -3)$.

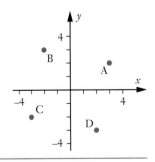

Data display

It is anticipated that you have already encountered the idea of counting or *tallying* data and organising it into tables, frequency tables and two way classification tables. The following forms of data presentation are also assumed to be familiar.

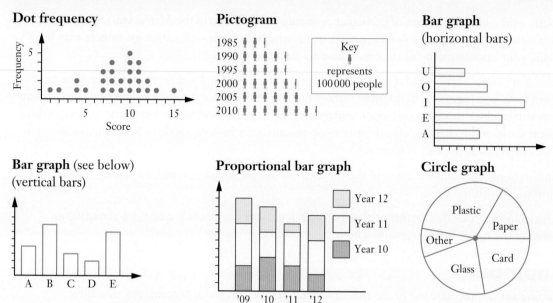

Dot frequency

Pictogram

Key
represents 100 000 people

Bar graph (horizontal bars)

Bar graph (see below) (vertical bars)

Proportional bar graph

Circle graph

Some texts call a bar graph with vertical bars a column graph and restrict the name 'bar graph' to those with horizontal bars. In this text we will not make that distinction and will refer to both as bar graphs.

Stem and leaf plot

```
    1 |14| 1 5 5
  5 1 |15| 0 3 8 3
3 2 8 |16| 3 6 7
  0 7 |17| 8 6
    6 |18| 5
```

Frequency histogram

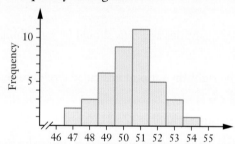

Data analysis

The **mean**, the **median** and the **mode** are all measures used to summarise a set of scores. They are all ways of giving an *average* or *typical* score for the set.

The **mean** is found by summing the scores and then dividing by the number of scores there are. The mean is the arithmetic average of the scores.

We use the symbol $\bar{x}$ to represent the mean of a set of scores.

For example, for the ten numbers: 19, 25, 29, 28, 23, 15, 27, 22, 24, 21

$$\bar{x} = \frac{19 + 25 + 29 + 28 + 23 + 15 + 27 + 22 + 24 + 21}{10}$$

The mean of the ten numbers is 23.3.

ISBN 9780170390262

The mean is a very useful measure of central tendency and is frequently used when analysing data. One disadvantage of the mean is that it can be greatly influenced by extreme scores, called **outliers**. For example if we add an eleventh score of 97 to the previous list the mean jumps from 23.3 to 30, i.e. the mean now exceeds all of the original ten scores. Clearly the score of 97 is a long way from the other scores. It is an extreme value and alters the mean significantly.

The **median** is the middle score in an ordered set of scores. If there are an even number of scores we say that the median is the mean of the 'middle two' in the ordered set.

For example, to determine the median of the seven numbers

$$29, 13, 27, 18, 33, 16, 29$$

we write them in order and choose the middle one:

$$13, 16, 18, \boxed{27}, 29, 29, 33$$

The median of the seven scores is 27.

To determine the median of the eight numbers

$$29, 13, 27, 18, 33, 16, 29, 13$$

we write them in order and find the mean of the middle two:

$$13, 13, 16, \boxed{18, \ 27}, 29, 29, 33$$

The median of the eight scores is 22.5.

The **mode**, or modal score, in a set of scores is the one that appears most frequently. There is of course no guarantee that the mode represents a score that is anywhere near the middle of the set of scores. It can be a useful and informative measure but is not necessarily 'central'. The mode is used when we want the 'most popular' value.

If there are two scores that are equally 'most popular' we say the set of scores is **bimodal**, because it has two modes. We do **not** find the mean of the two modes. (Sets of scores with more than two modes could be referred to as multimodal.)

For the ten numbers: 7, 8, 5, 9, 9, 11, 9, 11, 8, 5 the mode is 9.

The set of ten numbers: 5, 8, 5, 9, 9, 11, 9, 11, 8, 5 is bimodal.
                    The modes are 5 and 9.

Make sure you agree with the given mean, median and mode for the following set of numbers:

$$3, 3, 1, 4, 3, 0, 5, 5, 5, 3, 1, 4, 5, 4, 2.$$

The mean of the set of numbers is 3.2, the median is 3 and the set is bimodal with modes of 3 and 5.

As well as wanting to summarise the data using a mean or median or mode we may also be interested to know how widely spread the data is.

One way of indicating this is by stating the **range**, which is simply the difference between the highest score and the lowest score.

For example the set of numbers 7, 8, 5, 9, 9, 11, 9, 11, 8, 5 have a range of 6, obtained by working out (11 – 5).

Whilst the range is easy to calculate it is determined using just two of the scores and does not take any of the other scores into account. For this reason it is of limited use.

Formulae

From previous work, probably from *Unit One* of the *Mathematics Applications* course, you should be familiar with the idea of using a formula to determine the value of a variable, or pronumeral, that appears in the formula by itself and on one side of the equals sign, given the values of the variables, or pronumerals appearing on the other side.

For example:

Given $A = P + I$ we could determine A, given P and I:

If $P = 200$ and $I = 15$ it follows that

$$A = 200 + 15$$
$$= 215$$

Given $C = 2\pi r$ we could determine C, knowing π and given r:

If $r = 4$, it follows that

$$C = 2 \times \pi \times 4$$
$$= 25.133 \text{ (rounded to three decimal places)}$$

Given $s = ut + \frac{1}{2} at^2$ we could determine s, given u, a and t:

If $u = 4$, $a = 10$ and $t = 6$ it follows that

$$s = 4 \times 6 + \frac{1}{2} \times 10 \times 6^2$$
$$= 204$$

Algebra

You should also be familiar with evaluating **expressions** such as $2x + 3$, $5x - 2$, $5y + 4$, $2(x + 3)$, $3xy + 2z$ etc, given the values of x, y, and z.

Substitution code puzzle

If $x = 2$, $y = 3$ and $z = -5$ then $2x + 3 = 7$

$$5x - 2 = 8$$
$$5y + 4 = 19$$
$$2(x + 3) = 10$$
$$3xy + 2z = 8$$

The idea of **expanding brackets** should also be familiar to you:

The expression $3(x + 2)$ means 'three lots of $(x + 2)$'. Think of the bracket as a parcel containing an x and a 2. If we open the three parcels we have three xs and three 2s, i.e. $3x + 6$. We call this *expanding* the brackets.

Thus $5(x + 4)$ expands to $5x + 20$

          $7(2x + 5)$ expands to $14x + 35$

        $-2(3x - 4)$ expands to $-6x + 8$

Collecting like terms

If we are expanding several brackets we may be able to simplify our answer.

For example $3(2x + 1) + 5(x + 3) = 6x + 3 + 5x + 15$
$$= 11x + 18$$

Multiplying and dividing

For example $4(x + 3) - 3(x + 2) = 4x + 12 - 3x - 6$
$$= 1x + 6 \qquad \text{Usually written: } x + 6.$$

Expanding algebra

Similar triangles

To know whether two triangles are similar we can:

- see if the 3 angles of one triangle are equal to the 3 angles of the other triangle.

OR • see if the lengths of corresponding sides are in the same ratio.

OR • see if the lengths of two pairs of corresponding sides are in the same ratio and the angles between the sides are equal.

Once we know that two triangles are similar then we know that corresponding sides are in the same ratio.

The Pythagorean theorem

From unit one of this course you should be familiar with the following fact:

The square of the length of the hypotenuse of a right angled triangle is equal to the sum of the squares of the lengths of the other two sides.

Thus, for the triangle shown on the right, $c^2 = a^2 + b^2$

The theorem of Pythagoras allows us to determine the length of one side of a right triangle, knowing the lengths of the other two sides.

Probability

The work of chapter 13 assumes a basic understanding of the idea that the probability of something happening is a measure of the likelihood of it happening and this measure is given as a number between zero (no chance of happening) to 1 (certain to happen).

If we roll a normal die once the probability of getting a 3 is one sixth.

We write this as: $P(3) = \dfrac{1}{6}$.

For each spin of the spinner shown on the right

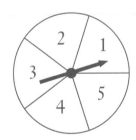

$$P(1) = \frac{1}{5}, \qquad P(2) = \frac{1}{5},$$

$$P(\text{odd number}) = \frac{3}{5}, \qquad P(\text{even number}) = \frac{2}{5},$$

$$P(\text{number} > 5) = 0, \qquad P(\text{number} < 6) = 1.$$

Geometry

It is assumed that you are familiar with the fact that when two straight lines intersect, the vertically opposite angles are equal, and with angle facts relating to alternate angles, corresponding angles and co-interior angles with parallel lines.

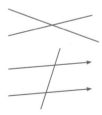

Use of technology to process and display data

Students following this course may come to this unit with very varied abilities in the use of technology such as graphic calculators and computers. Whatever your initial ability you are encouraged to make use of such technology whenever appropriate.

As the course progresses, do try to:

a gain familiarity with entering data into the columns of a calculator with statistical capabilities and showing various statistical information and displays about that data.

b become familiar with entering data into the columns of a spreadsheet on a computer or calculator and of carrying out straightforward operations on those entries such as adding a list of numbers, finding their average and presenting the data as a graph, as shown below.

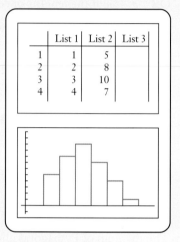

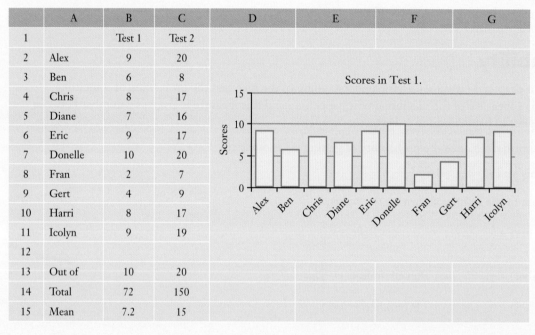

	A	B	C	D	E	F	G
1		Test 1	Test 2				
2	Alex	9	20				
3	Ben	6	8				
4	Chris	8	17				
5	Diane	7	16				
6	Eric	9	17				
7	Donelle	10	20				
8	Fran	2	7				
9	Gert	4	9				
10	Harri	8	17				
11	Icolyn	9	19				
12							
13	Out of	10	20				
14	Total	72	150				
15	Mean	7.2	15				

Scores in Test 1.

1.

Univariate data: Classify, organise and display

- Types of data
- Variables
- Categorical variables
- Displaying categorical data
- Numerical variables
- Displaying numerical data
- Histograms and bar charts
- Miscellaneous exercise one

Situation: Cardiovascular disease

A doctor is to give a lecture on cardiovascular disease, i.e. diseases of the heart or blood vessels, including heart attack and stroke. Using statistics for the previous year she wants to start her lecture by saying:

'Did you know that of the _____ deaths in Australia last year, from all causes,
_____% of these were due to cardiovascular disease. This means that last year,
on average, one Australian died from cardiovascular disease every _____ minutes.'

Copy and complete the above introduction to her talk using the information given in the table below to fill in the blanks. (Give the percentage to the nearest percent and the time to the nearest minute.)

TOTAL AUSTRALIAN DEATHS FOR YEAR PRIOR TO TALK: All ages

CAUSE OF DEATH	MALE	FEMALE	PERSONS
CARDIOVASCULAR DISEASE			
Coronary heart disease	12 433	11 137	23 570
Stroke	4 668	6 845	11 513
Other cardiovascular	4 856	6 195	11 051
(Sub-total)	21 957	24 177	46 134
CANCERS	22 039	17 183	39 222
TRANSPORT ACCIDENTS	1 224	414	1 638
ALL OTHER	22 021	21 699	43 720
ALL CAUSES	67 241	63 473	130 714

[Source of data: National Heart Foundation of Australia and The Australian Bureau of Statistics.]

The situation above involved you in making sense of information, or **data**, that was presented as a table, extracting relevant information from that table, and summarising that data in terms of percentages and times.

The data involved in the table would have been collected from government data bases where the cause of death for each deceased person, as stated on the death certificate, would be stored.

Data collection is often carried out to investigate some aspect of our lives. The methods by which we collect that data can vary as can the types of data we collect. The initial chapters of this text consider

- the various types of data,
- organising and displaying data,
- describing and interpreting data,
- comparing sets of data.

Types of data

For some investigations we collect the data ourselves by asking questions, by measuring, by experiment etc. This is called **primary data** – data we have collected ourselves.

Sometimes it is appropriate to use the data already collected by others, as in the situation on the previous page. For us this would be **secondary data** – data collected by others.

> Note: Internet access provides us with a ready supply of information regarding all sorts of subjects. Whilst much of this may well be valid we need to be cautious before assuming certain things about the information. We might consider such things as:
>
> - Was the method of data collection appropriate?
>
> - Is the information fairly presented?
>
> - Is the data collected from everyone in a particular situation or was sampling involved?
>
> - Are any summary statements correct?
>
> - Are any conclusions reasonable?
>
> One way to have confidence in the validity of the information is to use data from a reputable source. *The Australian Bureau of Statistics* (ABS), for example, would be one such credible source, as would other Government departments. Such sources are likely to have correctly considered collection methods, suitable presentation of data, appropriate conclusions etc.

Variables

If we ask someone what their favourite colour is or how tall they are the answers we get will vary from one person to another. Not everyone has blue as their favourite colour, not everyone is 163 cm tall etc. The responses will vary. Favourite colour and height are examples of **variables**.

If we are considering *one* variable, such as favourite colour, any data we collect will be **univariate**. However if we wanted to investigate how favourite colour may change with the age of a person we would be considering two variables – favourite colour and age. In such cases we would be considering **bivariate** data. This unit will only consider univariate data. Unit three of the *Mathematics Applications* course considers bivariate data.

Categorical variables

Consider the following questions that might be asked in some data collection activities:

- How do you get to school (or work) – walk, cycle, car, bus, train, other?

- Are you male or female?

- What is your favourite colour?

- Have you ever lived in a country other than Australia?

- How would you rank your fitness level: Low, Medium or High?

- What would you consider the best description of the current outside temperature – cold, cool, warm, or hot?

- What is your house number?

- What number are you in your rugby team?

Each of these questions allow us to place the person who responds into a particular **category** or **group**. The response might allow us to place the respondent in the group of males, or the group of people whose favourite colour is red, those who regard their fitness level as high, or who live in houses numbered eleven, or play number eight in their rugby team etc. The variable concerned, be it mode of transport, gender, favourite colour etc are all examples of **categorical variables**. Data associated with a categorical variable is called **categorical data**.

Notice that the first four questions each involve categories for which order is irrelevant. In general it would be pointless to suggest an order to the modes of transport, or to the gender categories, or place red ahead of blue as a category. We may rank order according to the numbers of people in that category but not on the basis of the category *names* themselves which have no natural order about them. Such categorical variables are called **nominal categorical variables**. (Name – nominal).

However the last four questions each involve categories that do have a natural *order* about them. These last four questions involve **ordinal categorical variables**, i.e. variables that do have a natural order about them. (Order – ordinal)

Notice though that whilst categorical variables can have categories that have numbers assigned to them these numbers are simply labels, they have no numerical significance, as in the responses to the last two questions in the above list.

Houses numbered 11, for example, are not necessarily bigger, better or more expensive than houses numbered 10. The house number is simply a label.

The number 8 player in one rugby team is not necessarily bigger, faster or better than the number 6 in another team. The number is simply a label indicating a position played and has no numerical significance.

Categorical variables	
Nominal Based purely on the names of the various categories, no order is suggested.	**Ordinal** The names of the categories, even though only labels, do suggest an order.

Displaying categorical data

We tend to use tables, pie charts and bar graphs to organise and display categorical data, as shown below.

State or Territory	Population at end of June quarter 2012 (nearest hundred)
Australian Capital Territory (ACT)	374 700
New South Wales (NSW)	7 290 300
Northern Territory (NT)	234 800
Queensland (Qld)	4 560 100
South Australia (SA)	1 654 800
Tasmania (Tas)	512 000
Victoria (Vic)	5 623 500
Western Australia (WA)	2 430 300
Total	22 680 500

Source: Australian Bureau of Statistics.

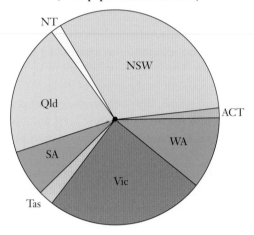

Pie Chart of Australian State and Territory populations as at end of June Quarter 2012. (Total population 22 680 500.)

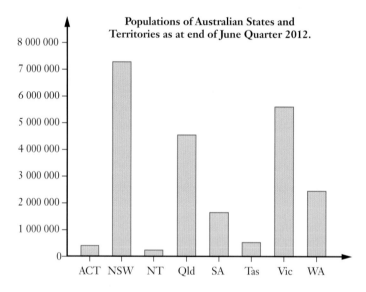

Populations of Australian States and Territories as at end of June Quarter 2012.

Notice that we may choose some particular order in which to display this nominal categorical data, for example in alphabetical order (as used above), or perhaps in order of population size or land area, or perhaps in order of state or territory formation date, or in order of number of letters in their non-abbreviated titles (!), etc, but on the basis of the categories themselves they have no natural order about them.

ISBN 9780170390262

Exercise 1A

For 1 to 12 classify each of the categorical variables given as either *nominal* or *ordinal*.

1 Country of birth.

2 Preferred writing hand: Left, Right or no preference.

3 Playing number in a basketball team.

4 Home telephone number.

5 Blood group: A, B, AB or O.

6 The number of the bus used to get to school.

7 A rating on sporting ability: Low, medium or high.

8 The construction type of a house: Brick, concrete, steel, timber, or other.

9 The state or territory of Australia that a person lives in.

10 Ow zumwon rayts ther spellin ablity: Poor, Okay, Good, Very Good, Excellent.

11 How someone voted in the last election.

12 The type of crop grown: Wheat, Barley, Oats etc.

13 OFFENCES AGAINST PROPERTY

In one year in Western Australia there were 197 117 offences categorised as an *offence against property*. These offences were further categorised as one of:

- Burglary: 38 410 offences
- Theft: 81 724 offences
- Fraud: 8552 offences
- Graffiti: 13 762 offences

- Stealing motor vehicle: 7618 offences
- Receiving: 1547 offences
- Arson: 1269 offences
- Property damage: 44 235 offences

[Source of data: Western Australia Police.]

Display this information both as a bar chart and as a pie chart and comment on the advantages and disadvantages of each form of display.

14 HOSPITAL BEDS

In one particular year the number of hospital beds available for patient care in each state or territory of Australia were as follows:

NSW	Vic	Qld	WA	SA	Tas	NT	ACT
27 543	22 502	16 623	8138	8021	2737	764	1083

The fact that NSW had the largest number of beds available should come as no surprise because it had the largest population. To be able to compare these numbers more meaningfully we need to consider them with respect to the population of each state or territory which, in that particular year was as follows:

NSW	Vic	Qld	WA	SA	Tas	NT	ACT
5 902 925	4 417 821	2 966 696	1 637 072	1 447 118	467 388	166 823	289 344

[Based on data from the Australian Bureau of Statistics.]

For each state or territory calculate the number of beds per 10 000 of population for this particular year, giving your answers correct to the nearest whole number, and display your answers as a bar graph.

RESEARCH

The figures in **Question 14** were actually for the early 1990s. Try to find more recent data for the number of beds per 10 000 of population in each state and territory and comment on how it compares with the data given above.

Numerical variables

Consider the following questions that might be asked in some data collection activities:

- How many people live in your house?
- How many people in your Mathematics class?
- How many pets do you have?
- How many of your teeth have been filled in some way?
- How tall are you?
- How far have you walked today?
- What is your blood pressure reading?
- What weight are you?

The responses to each of these questions will be **numerical** and the number will not just be a label, it will indicate a size or an amount. Some form of counting or measuring will be required to be done, or to have been done, for the response to be given. In each case the variable concerned, be it the number of people in a house, how many pets you have, your height, your blood pressure reading etc, are all examples of **numerical variables**. Data associated with a numerical variable is called **numerical data**.

Notice that the first four questions each involve responses that can only take specific values, in this case integer values. We cannot have 2.3 people living in a house, we cannot have 28.6 people in a Mathematics class. Numerical variables that can only take integer values are called **discrete variables**. Data associated with discrete variables is **discrete data**.

However the last four questions each involve responses that can take any value (usually within some realistic range). We can be 164.5 cm tall, we can walk 5.278 km in a day. Responses no longer have to be integer values and in practice the limit on the values taken are those of reasonableness (eg we cannot have 18.24 metres for the height of someone) and the accuracy of the measurement instrument used. Numerical variables that can take any value in an interval are called **continuous variables**. Data associated with a continuous variable is **continuous data**.

Generally, if counting is involved we have a discrete variable, if measurement is involved we have a continuous variable.

Exercise 1B

For 1 to 12 classify each of the numerical variables given as either *discrete* or *continuous*.

1 The number of rooms in a house.

2 The area of the block of land a house stands on.

3 The number of brothers and sisters a person has.

4 The length of a person's handspan.

5 The time it takes to get to school.

6 The lifetime of a rechargeable battery before it needs recharging.

ISBN 9780170390262

7 The number of car thefts in Australia in a week.

8 The temperature in degrees Celsius.

9 The number of people visiting a supermarket in a day.

10 The length of a car.

11 The capacity of a swimming pool.

12 The weight of a frog.

Displaying numerical data

In an attempt to analyse the use of its teachers a school noted the number of students in each of 20 year eleven classes. The school did not allow classes to run with more than 25 students and the smallest group had just 4 students. The number of students in the 20 classes gave rise to the following **dot frequency diagram**:

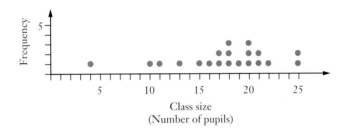

The variable quantity here, i.e. the number of students in each class, is an example of a **discrete** variable because the numerical variable can only take particular values, in this case integer values, from the given low of 4 to the high of 25.

Consider instead the continuous variable of the length of a new born baby and suppose that the lengths of 50 babies, recorded to the nearest centimetre, gave rise to the following table:

Length (cm)	47	48	49	50	51	52	53	54
Frequency	1	3	5	9	14	12	4	2

We could display this information as a dot frequency diagram, as shown on the right.

However it would be better to acknowledge the continuous nature of the data. The two babies recorded as 54 cm in length could have been anywhere from 53.5 cm to 54.5 cm. We can show this if we display the data as a **frequency histogram**. This is similar to a bar graph but always has frequency on the vertical axis, an ordered numerical scale on the horizontal axis and no gaps between the bars. Such a histogram for the data is shown on the next page.

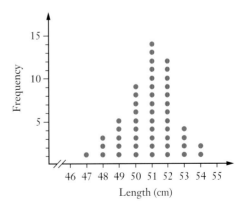

There was 1 baby recorded as 47 cm, when measured to the nearest centimetre. Thus for the 47 cm class the lower class boundary is 46.5 cm and the upper class boundary is 47.5 cm. This gives the first column in the histogram shown.

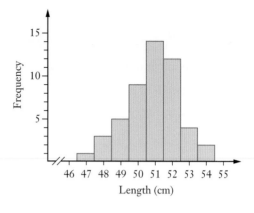

Note

- A histogram does not have gaps between the columns because, with continuous data, each class begins where the previous one leaves off. (Except by the apparent 'gap' between columns if a column has a frequency of zero.)

- We would usually have about 6 to 10 intervals. Too many and the table can be unmanageable, too few causes unnecessary bunching.

- It is usual to have class intervals of equal width. Intervals of different width could be misleading (and will be avoided in this text).

- The vertical axis on a histogram is always frequency.

- The horizontal axis on a histogram is a number line.

- Sometimes the frequency in each class may be shown as a fraction of the whole (*relative frequency*) or as a percentage of the whole (*percentage frequency*).

  Note that one of the histograms below also shows a **frequency polygon**, formed by connecting the middle of the top of each bar to the next, thus making the overall shape and continuity more evident.

Length (cm)	47	48	49	50	51	52	53	54
Frequency	1	3	5	9	14	12	4	2
Relative frequency	0.02	0.06	0.10	0.18	0.28	0.24	0.08	0.04
Percentage frequency	2%	6%	10%	18%	28%	24%	8%	4%

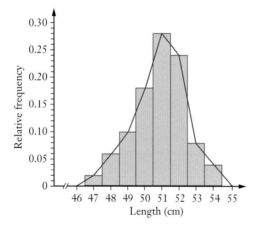

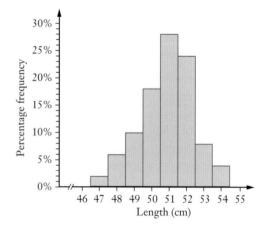

ISBN 9780170390262

- Whilst we might expect that discrete data would not be shown as a histogram because there would be gaps between the columns, histograms are such a convenient form of representation that they are frequently used to display discrete data. In such cases we choose our class boundaries to be midway between the possible discrete values. For example the discrete data shown below left as a dot frequency diagram can be displayed as a frequency histogram, as shown below right.

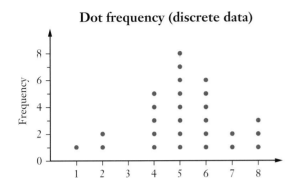

Dot frequency (discrete data)

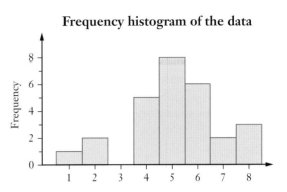

Frequency histogram of the data

- Histograms can be a useful form of display when we have discrete data that is *grouped* for convenience. For example, consider the following 25 scores:

$$35 \quad 46 \quad 12 \quad 34 \quad 18 \quad 20 \quad 25 \quad 24 \quad 11$$
$$14 \quad 29 \quad 9 \quad 27 \quad 23 \quad 32 \quad 38 \quad 30 \quad 17$$
$$22 \quad 19 \quad 36 \quad 28 \quad 33 \quad 4 \quad 21$$

No scores are repeated so if we were to display the scores as a frequency table, or as a dot frequency graph, we would have 25 scores shown, each with a frequency of 1. In this case the data may be better presented grouped. Using the intervals 1–5, 6–10, 11–15, 16–20 etc., the grouping becomes:

Score	1–5	6–10	11–15	16–20	21–25	26–30	31–35	36–40	41–45	46–50
Frequency	1	1	3	4	5	4	4	2	0	1

Some information is now lost because the 25 scores themselves are no longer given but the grouping can make the overall distribution more evident. The grouped data could be displayed as a histogram.

Histograms and bar charts

Both bar charts and histograms can show the frequency of something occurring, so what is the difference between them? One important difference is that histograms show a normal number line on the horizontal axis, bar graphs show categories. This means that the bars of a bar graph are not bound by order and can be moved around. We might for example arrange the bars in order of increasing height. The bars in a histogram cannot be moved around. They must be presented in order, with the horizontal axis giving this order. Further, the beginning of one category in categorical data does not logically take over from the end of another. If we are drawing a bar graph about the pets people have, our categories might be cat, dog, horse, etc. The categories are quite separate – a dog does not start where a cat leaves off. Hence bar charts tend to have gaps between the bars. However with ordered numerical data, especially of a continuous nature, one number interval does indeed take over where the previous one left off. Hence we put no gaps between the bars of a histogram.

Histograms

EXAMPLE 1

Sixty self-employed bricklayers were surveyed regarding the number of hours they worked during one particular week (to the nearest hour). The results of the survey are shown.

Display the data as a histogram.

Number of hours	Frequency (i.e. No. of bricklayers)
10 → 19	2
20 → 29	5
30 → 39	20
40 → 49	24
50 → 59	8
60 → 69	1

Solution

The first column of the histogram will extend from 9.5 to 19.5, the second column from 19.5 to 29.5 and so on. The completed histogram is shown below.

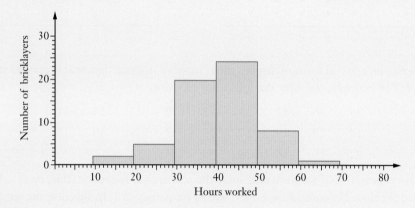

EXAMPLE 2

The road accident statistics for a country for one year showed that for motorcyclists (drivers not passengers) in the age range fifteen to fifty-nine, 186 had died in road accidents with the distribution of the ages of these riders as shown on the right.

Display this information as a frequency histogram.

Age (x yrs)	Drivers killed
$15 \leq x < 20$	40
$20 \leq x < 25$	59
$25 \leq x < 30$	29
$30 \leq x < 35$	19
$35 \leq x < 40$	16
$40 \leq x < 45$	11
$45 \leq x < 50$	8
$50 \leq x < 55$	2
$55 \leq x < 60$	2

ISBN 9780170390262

Solution

The first column of the histogram will extend from 15 to 20, the second column from 20 to 25 and so on. The completed histogram is shown on the right.

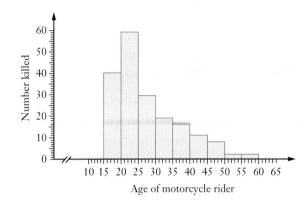

Exercise 1C

1 A normal, fair, six-sided die is rolled 72 times and the number displayed on the uppermost face is noted each time.

Given that one of the three frequency histograms shown below displays the results obtained for these 72 rolls which of the three is it most likely to be?

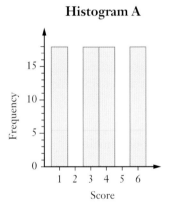

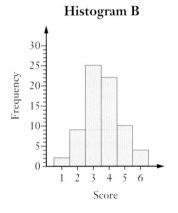

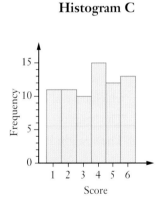

2 The thirty students in a class measured their heights and the results are displayed in one of the three histograms shown below. Which one is it most likely to be?

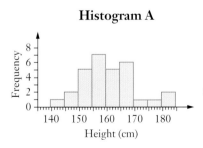

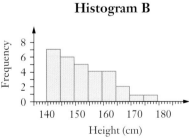

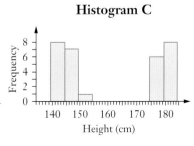

3 Two normal fair six-sided dice are rolled and the numbers displayed on the two uppermost faces are added together and the total noted. This process is carried out 72 times.

Given that one of the three frequency histograms shown below displays the results obtained for this activity which of the three is it most likely to be?

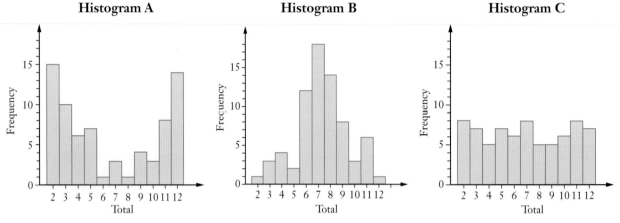

Histogram A	Histogram B	Histogram C

4 What shape will the histogram be?

Sketch what you consider to be a reasonable histogram for each of the following. Your sketch should show what you consider to be a reasonable shape for the histogram to have. There is no need to include any numbers on the axes of your sketch graphs except for parts **a** and **b** which should have numbers on the horizontal axis.

a Rolling a fair 8-sided die approximately one hundred times.

b The number of children in approximately one hundred randomly selected families, each of which have at least one child.

c The heights of a large number of adult males.

d The straight line distance from home to school for students at your school.

e The lengths of new-born babies.

For at least some of Questions 5 to 10 generate the required histogram using a computer or calculator.

5 HORTICULTURE

Some seeds were planted and, some weeks later, the heights of the seedlings were measured and recorded, to the nearest centimetre. The results were as follows:

Length (cm)	6	7	8	9	10	11	12	13	14	15
Frequency	2	4	10	15	14	11	9	7	3	1

Display this information as a frequency histogram.

ISBN 9780170390262

6 METEOROLOGY

The maximum temperature recorded at Perth airport for each day of December in one particular year gave rise to the following data:

Maximum temperature (°C)	20 → 24	25 → 29	30 → 34	35 → 39	40 → 44
Frequency (number of days)	3	13	8	6	1

[Source: Bureau of Meteorology.]

Display this information as a frequency histogram.

7 ROAD ACCIDENTS

The road accident statistics for a country for one year showed that for drivers in the age range fifteen to fifty-nine, 708 had died in road accidents. The distribution of the ages (x years) of these drivers are shown in the table.

Display this information as a frequency histogram.

Age (x years)	Drivers killed
$15 \leq x < 20$	138
$20 \leq x < 25$	131
$25 \leq x < 30$	75
$30 \leq x < 35$	95
$35 \leq x < 40$	79
$40 \leq x < 45$	71
$45 \leq x < 50$	57
$50 \leq x < 55$	39
$55 \leq x < 60$	23
Total	708

8 TIME ESTIMATION

One hundred and fifty students were asked to estimate a time period of one minute. The time periods they thought were one minute were actually the following number of seconds, to the nearest second:

31	68	46	66	54	48	70	60	62	48	97	53	50	56	60
52	56	92	50	43	65	45	80	53	64	56	67	59	41	49
65	75	50	51	66	75	50	56	40	57	64	44	69	71	51
51	64	89	74	49	54	57	67	54	59	47	79	51	54	50
59	90	49	61	52	64	77	46	74	48	66	49	76	66	41
43	50	81	62	68	44	49	66	52	84	45	84	42	52	59
65	74	42	73	54	50	73	60	49	60	54	52	69	56	50
62	47	51	45	50	67	59	38	65	46	56	85	48	54	51
50	67	65	54	65	48	51	54	62	52	51	53	70	43	57
47	64	54	69	43	86	62	69	51	64	52	76	64	68	46

Arrange the data as a grouped frequency table with classes of equal width as follows:
30 – 39, 40 – 49, etc., up to 90 – 99.

Display the grouped data as a frequency histogram.

9 WEIGHT

200 males and 200 females, all aged between 30 and 40, took part in a survey which recorded (amongst other things) the weight of each person, recorded to the nearest kg. The results were as follows:

MALES			FEMALES	
Weight (kg)	Frequency		Weight (kg)	Frequency
30 – 39	0		30 – 39	5
40 – 49	2		40 – 49	12
50 – 59	10		50 – 59	73
60 – 69	35		60 – 69	66
70 – 79	68		70 – 79	27
80 – 89	53		80 – 89	8
90 – 99	21		90 – 99	5
100 – 109	9		100 – 109	3
110 – 119	2		110 – 119	1
Total	200		Total	200

Display these results as two separate frequency histograms and include the frequency polygon on each one.

10 HEIGHT

50 males and 50 females, all aged between 20 and 30, took part in a survey which recorded (amongst other things) the height of each person, recorded to the nearest cm. The results were as follows:

MALES			FEMALES	
Height (cm)	Frequency		Height (cm)	Frequency
140 – 149	0		140 – 149	1
150 – 159	1		150 – 159	11
160 – 169	6		160 – 169	28
170 – 179	26		170 – 179	8
180 – 189	14		180 – 189	2
190 – 199	2		190 – 199	0
200 – 209	1		200 – 209	0

Display these results as two separate **percentage** frequency histograms.

ISBN 9780170390262

Miscellaneous exercise one

This miscellaneous exercise may include questions involving the work of this chapter and the ideas mentioned in the Preliminary work section at the beginning of the book.

1 State true or false for each of the following statements:

  a $3^2 = 6$ **b** $(-3)^2 = -9$ **c** $-5 - 3 = 8$

  d $5^2 + 3^2 = 8^2$ **e** $5 + 3 \times 2 = 16$ **f** $3(x + 1) = 3x + 1$

2 Classify each of the variables mentioned in parts (a) to (h) as one of:

Nominal categorical, Ordinal categorical, Discrete numerical, Continuous numerical

  a   Favourite soccer team.

  b   Waist measurement.

  c   Number of people in a car.

  d   Mode of transport: Walk, cycle, bus, train, tram, other.

  e   Interest in sport: Not at all, weak, medium, strong, full on.

  f   Nationality of mother.

  g   Distance from home to school.

  h   The number of peas in a pod.

Shutterstock.com/Jag_cz

3 Find the mean of each of the following sets:

  a   131, 120, 141, 122, 136.

  b   2.4, 3.7, 1.9, 0, 2.3, 3.2, 1.6, 1.7.

  c   27, 18, 31, 33, 39, 27, 41, 29, 21, 27.

4 Find the median of each of the following sets:

  a   15, 17, 21, 22, 23, 25, 25.

  b   19, 21, 13, 28, 22, 25, 19, 22, 17.

  c   10, 17, 11, 15, 23, 11, 21, 12, 17, 9.

5 Find the mode of each of the following sets:

  a   3, 2, 1, 4, 3, 0.

  b   11, 13, 17, 10, 13, 14, 17.

  c   22, 15, 21, 22, 18, 19, 16, 22, 17, 19.

6 The instructions for mixing a weedkiller says to mix concentrate and water in the ratio 1 : 300. How much water should be added to 25 millilitres of concentrate?

7 Find the mean, median, mode and range of each of the following sets:

  a   33, 37, 38, 40, 40.

  b   131, 93, 124, 107, 68, 131, 70, 110, 84.

  c   18, 15, 17, 18, 15, 18, 18, 17, 19, 17.

8 Expand each of the following and simplify where possible.

  a $3(5x - 2)$ **b** $4(7 - 2x)$ **c** $-3(2x + 7)$

  d $8(1 - 2x)$ **e** $5(2p - 7)$ **f** $3(2h - 5)$

  g $5 + 2(1 + 3x)$ **h** $4(2x + 1) - 5(3 + 2x)$ **i** $2x + 3(5 - 2x)$

  j $2(5 + q) - 3(1 - 2q)$ **k** $6(2w + 3) - 5w + 4$ **l** $2(p + 6) - 4(3 - p)$

9 The framework shown is to be made of steel. Find the total length of steel required for the framework giving your answer rounded up to the next whole metre.

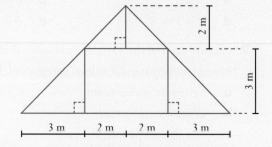

10 HOLIDAY ACTIVITIES

A sports club offers holiday activities for young people aged at least 5 but under 15. The enrolment forms required participants to give, amongst other things, their age in years and months. To determine which ages their programme of activities suited, club officials considered the ages of participants. These are shown below.

5y 3m	14y 5m	7y 2m	6y 2m	10y 5m	8y 10m	8y 3m	7y 5m
12y 2m	6y 1m	13y 3m	9y 9m	6y 1m	6y 3m	11y 5m	8y 5m
9y 11m	13y 0m	7y 10m	12y 2m	8y 0m	14y 7m	6y 9m	9y 11m
8y 7m	7y 1m	8y 0m	7y 5m	7y 0m	5y 3m	10y 0m	6y 7m
14y 10m	11y 2m	5y 8m	12y 1m	11y 7m	7y 3m	12y 3m	6y 11m
6y 1m	7y 9m	10y 2m	11y 2m	5y 2m	7y 3m	7y 4m	9y 2m
13y 2m	12y 4m	7y 2m	6y 3m	9y 4m	8y 0m	6y 9m	7y 7m
11y 11m	6y 1m	13y 9m	13y 7m	6y 9m	12y 7m	11y 2m	6y 7m
5y 1m	9y 0m	7y 7m	14y 0m	13y 7m	6y 9m	9y 7m	12y 11m
8y 7m	14y 1m	13y 9m	5y 11m	8y 11m	12y 7m	5y 7m	8y 1m
8y 6m	7y 11m	6y 4m	13y 10m	7y 6m	6y 5m	6y 2m	6y 6m

  a   Arrange these ages into a grouped frequency table as shown (tally shows first column entered):

  b   Display the grouped data as a frequency histogram.

Age (x years)	Tally	Frequency
$5 \leq x < 6$	II	
$6 \leq x < 7$	I	
$7 \leq x < 8$		
$8 \leq x < 9$	III	
$9 \leq x < 10$	I	
$10 \leq x < 11$		
$11 \leq x < 12$	I	
$12 \leq x < 13$	I	
$13 \leq x < 14$	I	
$14 \leq x < 15$	I	

2.

Summarising data and describing distributions

- Combining groups
- Use of statistical functions on a calculator
- Grouped data
- Describing a distribution of scores
- Miscellaneous exercise two

Situation: Company salaries

A particular company employs thirteen people:

1 managing director,

2 managers,

1 accountant,

3 chargehands,

6 machine operators.

The salary structure in the company is shown below.

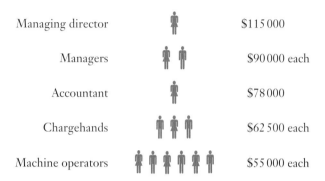

Managing director		$115 000
Managers		$90 000 each
Accountant		$78 000
Chargehands		$62 500 each
Machine operators		$55 000 each

- Find the company's mean salary, median salary and modal salary.

- If a group were arguing that higher salaries should be awarded to the people working for this company which of the previous three answers would they claim to be the average to best suit their argument?

- If the managing director wished to quote an average salary for this company, in support of her claim that the average salary was already high, which average best suits her argument?

- If this company took over another company, retaining all 7 employees of the other company on their existing salaries, which had a mean of $66 000, what would be the mean salary of the new '20 employee' company?

2. Summarising data and describing distributions ●●○●●●●●●●●●●

In the situation on the previous page you had to summarise information by determining the mean, median and mode, concepts that you were reminded of in the *Preliminary work* section at the beginning of this book and that some of the questions in *Miscellaneous exercise one* required you to find. The mean, median and mode summarise the location of a set of scores. They are *summary statistics* and are *measures of location*. The mean and the median give an indication of *central tendency*.

The next three examples remind you how to determine these quantities from a frequency table (example 1), a dot frequency graph (example 2) and a stem and leaf plot (example 3).

EXAMPLE 1

The fifty scores shown below

14	17	15	17	12	16	19	16	17	16
16	10	15	17	18	17	14	16	17	16
18	19	20	14	15	18	18	18	15	17
15	16	17	18	16	16	13	15	18	15
17	17	15	16	19	15	17	18	13	14

can be neatly displayed in the form of a frequency table, as shown below:

Score	10	11	12	13	14	15	16	17	18	19	20
Frequency	1	0	1	2	4	9	10	11	8	3	1

Determine the mode, the mean and the median of the fifty scores.

Solution

The mode is readily determined from this table by seeing which score has the greatest frequency. The mode of the set of scores is 17.

The table tells us that we have one 10, zero 11s, one 12, two 13s etc. Hence, to determine the total of the fifty scores we calculate:

$$
\begin{aligned}
& 1 \times 10 + 1 \times 12 + 2 \times 13 + 4 \times 14 + 9 \times 15 + 10 \times 16 + 11 \times 17 + 8 \times 18 + 3 \times 19 + 1 \times 20 \\
= \quad & 10 \quad + \quad 12 \quad + \quad 26 \quad + \quad 56 \quad + \quad 135 \quad + \quad 160 \quad + \quad 187 \quad + \quad 144 \quad + \quad 57 \quad + \quad 20 \\
= \quad & 807
\end{aligned}
$$

Hence the mean will be given by $\dfrac{807}{50}$ i.e. 16.14.

The median of fifty scores will be the mean of the 25th and 26th scores once the scores have been written in order. Summing the frequencies in the above table from the left end we see that the 25th and 26th scores will both be 16. Thus the median score is 16.

EXAMPLE 2

Determine the mode, mean, median and range for the set of scores shown in the dot frequency graph.

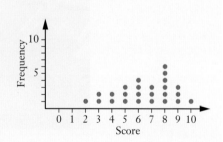

Solution

The score which occurs more frequently than any other is 8. Hence the mode = 8

The dot frequency graph tells us that we have one score of 2, two scores of 3 etc.

Hence the mean = $\dfrac{1\times2+2\times3+2\times4+3\times5+4\times6+3\times7+6\times8+3\times9+1\times10}{25} = 6.44$

The median of 25 numbers will have twelve numbers either side of it, i.e. it will be the thirteenth number. Counting along the dot frequency graph we see that the thirteenth number will be a seven. Hence the median is 7.

Remember that the range is the difference between the highest score (10) and the lowest score (2). Hence the range is 8.

EXAMPLE 3

Stem-and-leaf plots

Determine the mode, median, mean and range for the set of twenty scores shown in the stem and leaf diagram on the right.

```
12 | 7
13 | 9  3  1
14 | 2  0  3  2  7
15 | 2  5  5  1  9  5
16 | 8  3  8
17 | 3  0
```

Solution

By inspection the mode is 155.

With twenty scores the median will be between the 10th and 11th scores, when the scores are considered in order of size (see right). Hence the median will be the mean of 151 and 152, i.e. 151.5.

By calculation, the mean is 150.65 and the range is 173 – 127, i.e. 46.

```
12 | 7
13 | 1  3  9
14 | 0  2  2  3  7
15 | (1  2) 5  5  5  9
16 | 3  8  8
17 | 0  3
```

Note: If you look for a formula for the mean of a set of numbers in books containing mathematical formulae you may not find the rule stated in the form:

$$\text{Mean} = \frac{\text{The sum of the scores}}{\text{The number of scores there are}}$$

Instead the formula may involve symbols like $\bar{x}$ and Σ, as explained below.

- As mentioned in the *Preliminary work*, we use the symbol $\bar{x}$ to indicate the mean of a set of scores. For the twenty numbers of example 3, $\bar{x} = 150.65$.
- The Greek letter Σ, pronounced sigma, is used in Mathematics to indicate that numbers are being added together, i.e. a *summation* is being determined. Thus if we consider the numbers 8, 7, 6, 11 to be values of x then $\Sigma x = 8 + 7 + 6 + 11 \ (= 32)$.
- Putting the above ideas together, for a set of n scores it follows that

$$\bar{x} = \frac{\Sigma x}{n}$$

If we use f to indicate the frequency with which each score occurs it further follows that for data given as a frequency table

$$\bar{x} = \frac{\Sigma fx}{\Sigma f}$$

Combining groups

Knowing the number of scores and their sum we can calculate the mean of the scores. For example if 15 scores have a sum of 108 the scores have a mean of

$$\frac{108}{15} = 7.2$$

It also follows that if we know the number of scores and their mean we can find the sum of the scores.

For example, knowing that 15 scores have a mean of 7.2 the scores must have a sum of

$$15 \times 7.2 = 108$$

Indeed you probably used this idea if you managed the last dot point of the *Situation* at the beginning of the chapter. In that situation we were told that the 7 employees of a company had a mean salary of $66 000 from which we could calculate the total amount paid out for these salaries as

$$7 \times \$66\,000 = \$462\,000$$

This idea of determining the total of a set of scores, knowing the mean and the number of scores can be useful when solving some problems as examples 4, 5 and 6 show.

EXAMPLE 4

The mean of six scores is 23.5. If five of the scores were 17, 20, 19, 25 and 30 find the sixth score.

Solution

If six scores have a mean of 23.5 then these six scores have a sum of $23.5 \times 6 = 141$.

The five given scores have a sum of $\qquad\qquad 17 + 20 + 19 + 25 + 30 = 111$

Thus the sixth score must be $\qquad\qquad\qquad\qquad 141 - 111 = 30$

The sixth score is 30.

EXAMPLE 5

To pass a particular course a student needs to gain a mean of at least 55% in the five tests that form the course assessment. In the first four tests the student achieves marks of 46%, 57%, 54% and 57%. What percentage mark must the student gain in test five if they are to pass the course?

Solution

To gain a 55% average in 5 tests the total marks in the 5 tests must be $55 \times 5 = 275$

The first 4 tests have a sum of $\qquad\qquad\qquad 46 + 57 + 54 + 57 = 214$

Thus in the fifth test the student needs $\qquad\qquad\qquad 275 - 214 = 61$

The student needs to score at least 61% in the fifth test in order to pass the course.

EXAMPLE 6

In a test the 15 girls in a class score a mean mark of 21.2 and the ten boys score a mean mark of 22.4. Calculate the mean for the whole group of 25 students.

Solution

The 15 girls achieved a mean of 21.2 thus they gained a total mark of $15 \times 21.2 = 318$.

The 10 boys achieved a mean of 22.4 thus they gained a total mark of $10 \times 22.4 = 224$.

Thus the 25 students gained a total mark of $\qquad\qquad 318 + 224 = 542$

The 25 students achieved a mean mark of $\qquad\qquad \dfrac{542}{25} = 21.68$

Exercise 2A

Find the mean, median, mode and range of each of the distributions shown in questions 1 to 6 (correct to one decimal place if necessary).

1

Score	0	1	2	3	4	5
Frequency	1	2	2	3	4	7

2

Score	0	1	2	3	4	5	6	7	8	9	10
Frequency	1	0	2	3	2	4	7	6	2	3	1

3

Score	15	16	17	18	19	20
Frequency	1	2	7	8	3	1

4

Score	98	100	101	103	104	105
Frequency	1	1	3	3	1	1

5

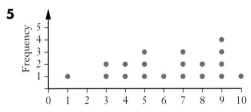

6

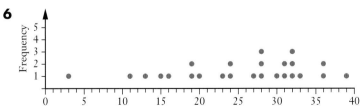

7 The stem and leaf diagram on the right can accommodate numbers from 110 to 169, the first two digits being the stem and the last digit the leaf. The diagram shows the scores recorded by 25 competitors in a shooting competition. For the scores recorded determine:

```
16 | 4 4
15 | 3 1 8 7
14 | 0 1 7 9 5 3
13 | 0 9 7 4 2 5 6 2
12 | 0 6 5
11 | 9 8
```

a the lowest score **b** the highest score

c the median score **d** the mean score

8 The stem and leaf diagram on the right shows scores achieved by 24 students in an exam that was marked out of 120. The stem is the tens digit and the leaf is the units digit. Determine

```
9 | 4 0 2
8 | 2 5 8 1 2 5 4 3
7 | 5 7 4 4 7
6 | 1 7 9 8 4 8 6
5 |
4 | 8
```

a the lowest score

b the highest score

c the median score

d the mean score (correct to 1 decimal place)

9 Estimate the mean number of thumbs per Australian adult.

10 The dot frequency diagram on the right shows the marks obtained by 30 year 8 students in a mental arithmetic test.

Determine the mean, median and mode for this distribution of marks.

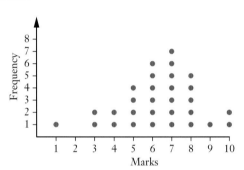

11 A class of school students calculated their mean age as 17.2 years, correct to one decimal place. This average did not include the teacher. If the teacher were to be included would the mean be increased or decreased?

12 The mean of eight scores is 54.25. If seven of the scores were 60, 50, 37, 60, 55, 32 and 65 find the eighth score.

13 Three mathematics classes sat the same exam. The mean marks for the classes were 55%, 62% and 56% and the number of students in each class were 24, 15 and 21 respectively. Find the mean for the three groups put together.

14 To pass a particular course a student has to achieve a mean of at least 50% in the ten pieces of work that form the assessment items. In the first nine of these pieces of work the student achieves a mean of 46%. What percentage mark must the student achieve in the tenth item if he is to pass the course?

15 The mean of 25 scores is 54. If 20 of the scores had a mean of 55 find the mean of the other five scores.

16 The 13 boys in a class gained a mean mark of 57% in a test in which the class mean was 59%. If the class consisted of 20 students altogether find the mean achieved by the girls in the class. (Give your answer correct to one decimal place.)

For each of the dot frequency graphs shown in questions 17 to 32, without actually calculating the mean and the median, state which one of the following statements apply:

- The mean is the same as the median.
- The mean is greater than the median.
- The mean is less than the median.

17

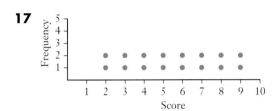

18

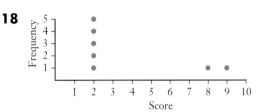

19

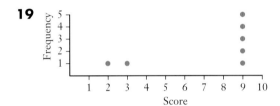

20

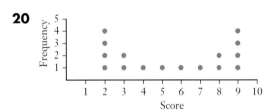

21

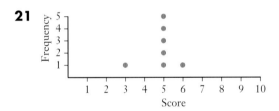

22

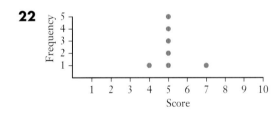

23

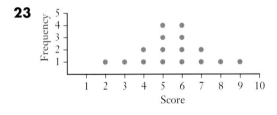

24

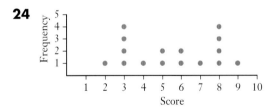

25

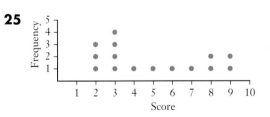

26

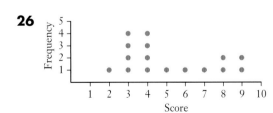

27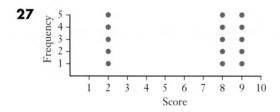

28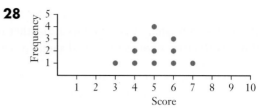

29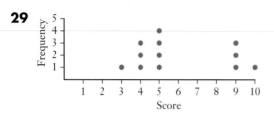

30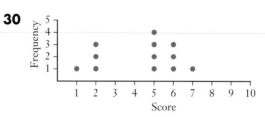

31

32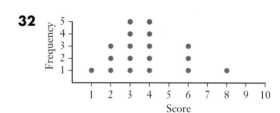

33 Whilst each of the dot frequency diagrams shown in this question feature 40 data points they show quite different distributions of the 40 scores.

Write a few sentences describing each distribution of 40 scores.

Some useful words and phrases that you might consider using in your descriptions could include:

lowest score	highest score	tightly packed	spread out
clusters	gaps	outliers	uniform
groups	dense regions	symmetrical	

a

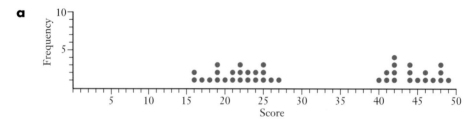

b

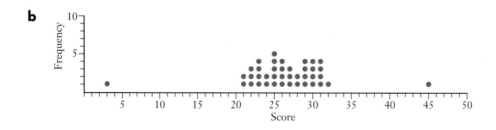

c

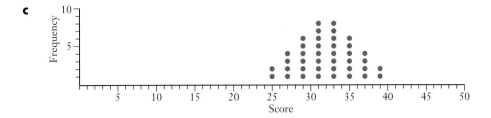

d

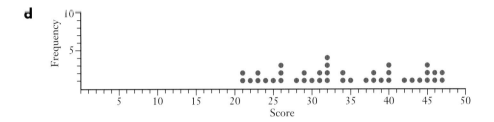

e

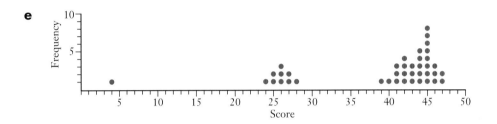

34 In each part of this question two dot frequency graphs are given, Graph A and Graph B. In each pair the two graphs involve sets of data with the same number of data points. However, in each case the two graphs are quite different. Write a few sentences comparing graphs A and B in each case. Whilst you do not need to determine the mean and the median exactly for any of the graphs your answers should include comparisons of means, medians and spread.

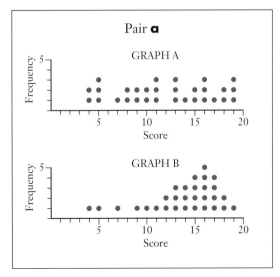

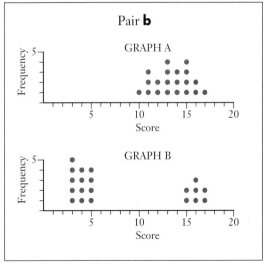

Use of statistical functions on a calculator

When asked to determine the mean of a set of scores are you adding up all of the scores and then dividing by the number of scores or are you putting the scores into a calculator and using its ability to determine the mean?

Many calculators have an inbuilt ability to output various summary statistics for data that is put into the calculator.

This usually involves the following steps:

1   Setting or selecting the statistics facility of your calculator.

2   Clearing any statistical data already in your calculator.

3   Inputting the data.

4   Outputting the required data.

Your calculator is able to output statistical information for the set of data you put in. The calculator may display more statistical information than you are familiar with at present but somewhere in the list it may well display the number of scores in the set, often listed as 'n', the mean of the scores, often listed as $\bar{x}$, the median, the mode and various other things, some of which you will encounter in later mathematical studies.

For example, below left shows such a display for the data set:

$$3, 5, 2, 0, 2$$

Scrolling down such a display shows more information as shown below right.

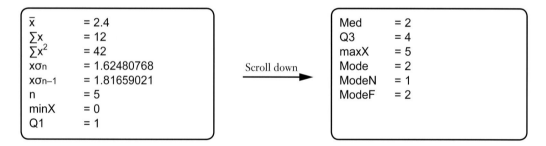

Can you locate   the mean,
                 the number of scores,
                 the median,
                 the mode?

. . . .

Get to know **your** calculator with regard to inputting data and getting it to display summary statistics for the data.

. . . .

ISBN 9780170390262

If the data is presented as a frequency table the information can be put into many calculators in this frequency form, i.e. we do not need to input all of the scores individually. Usually one column is used for the scores and another for the frequencies. We then set the calculator to read the information in each column appropriately.

For example, the following table

Score	10	11	12	13	14	15	16	17	18	19	20
Frequency	1	0	1	2	4	9	10	11	8	3	1

could be put into two columns, see display below left, and the summary statistics displayed, see below right (more statistics being available if we were to scroll down).

	List 1	List 2	List 3	List 4
1	10	1		
2	11	0		
3	12	1		
4	13	2		
5	14	4		
6	15	9		

$\bar{x}$	= 16.14
$\sum x$	= 807
$\sum x^2$	= 13205
$x\sigma n$	= 1.897472
$x\sigma n{-}1$	= 1.91673617
n	= 50
minX	= 10
Q1	= 15

Once again, get to know how to put information from a frequency table into your calculator and how to output the mean etc.

Grouped data

As was mentioned in chapter one, continuous numerical data is often grouped because of rounding and sometimes discrete numerical data is grouped for convenience. For example suppose we are given a set of fifty numbers with very few repeats. If we were to display the scores as a dot frequency graph, we would have almost as many 'columns' to our graph as we have scores as most of the scores occur just once:

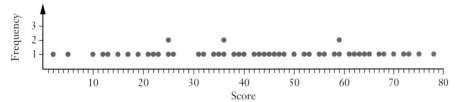

The data may be better presented in groups or class intervals:

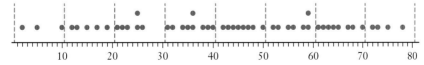

giving the grouped frequency table:

Score	1 → 10	11 → 20	21 → 30	31 → 40	41 → 50	51 → 60	61 → 70	71 → 80
Frequency	3	5	6	9	8	7	8	4

If we only have the grouped data and do not know the original scores we talk of the **modal group** or **modal class**, rather than referring to a mode. For the frequency table just encountered, and shown again below, the modal class is the $31 \rightarrow 40$ class.

Score	$1 \rightarrow 10$	$11 \rightarrow 20$	$21 \rightarrow 30$	$31 \rightarrow 40$	$41 \rightarrow 50$	$51 \rightarrow 60$	$61 \rightarrow 70$	$71 \rightarrow 80$
Frequency	3	5	6	9	8	7	8	4

Similarly we find the class in which the median lies, called the **median group** or **median class**, in this the case the $41 \rightarrow 50$ class.

To determine a mean we assume all the scores in an interval are at the **midpoint** of the interval. Clearly this is unlikely to be the case but it will give a reasonable estimate for the mean when a large number of scores are involved. Thus for the above table we calculate the mean based on three scores of 5.5, five scores of 15.5, six scores of 25.5 etc.

Check that you agree that applying this idea to the above table of grouped data gives an estimated mean of 42.9.

Exercise 2B (Use the statistical capability of your calculator.)

Find the mean, median, mode and range of each of the sets of data given in questions 1 to 7 (correct to one decimal place if necessary).

1 125, 137, 137, 143, 153, 162, 165.

2 85, 85, 78, 72, 83, 78, 90, 89, 78.

3 8, 34, 19, 14, 25, 15, 40, 26, 17, 30.

4 55, 42, 36, 63, 45, 35, 76, 50, 50, 58,
40, 72, 35, 80, 75, 66, 48, 35, 62, 35,
66, 40, 56, 52, 38.

5

Score	0	1	2	3	4	5	6	7
Frequency	8	12	18	20	9	3	0	1

6

Score	5	6	7	8	9	10	11	12
Frequency	24	35	17	28	33	31	27	19

7

Score	15	16	17	18	19	20
Frequency	1	3	5	13	16	22

In the next four questions use the midpoint of each class interval to determine the mean of each of the following distributions, correct to one decimal place.

8

Score	Frequency
1 → 5	15
6 → 10	28
11 → 15	7
16 → 20	3
21 → 25	1
26 → 30	1

9

Score	Frequency
1 → 4	3
5 → 8	8
9 → 12	15
13 → 16	8
17 → 20	3

10

Score	Frequency
20 → 24	6
25 → 29	10
30 → 34	17
35 → 39	7
40 → 44	5
45 → 49	4
50 → 54	1

11

Score (x)	Frequency
$0 \le x < 20$	5
$20 \le x < 40$	13
$40 \le x < 60$	21
$60 \le x < 80$	72
$80 \le x < 100$	54

12 As part of the process of assessing the value of a property a real estate agent considers the prices of other properties recently sold in the same area. The selling prices of ten such properties were as follows:

$437\,000 $425\,000 $456\,000 $421\,000 $442\,000
$445\,000 $441\,000 $437\,000 $432\,000 $540\,000

Find the mean and median of these prices.
How many of the ten prices are lower than the mean?
How many of the ten prices are lower than the median?

13 A real estate survey investigated the number of bedrooms in each of 100 houses in a particular area. The results were as shown below:

Number of bedrooms	1	2	3	4	5	6	7
Frequency	2	3	37	49	5	3	1

Find the mean number of bedrooms per house for these houses.

14 A company employs 25 people and has seven salary levels. The number of employees on each level are shown in the following table.

Salary	$62\,000	$68\,000	$71\,000	$78\,000	$85\,000	$100\,000	$110\,000
No. of employees	4	10	5	1	2	2	1

Calculate **a** the modal salary,
b the median salary,
c the mean salary.

15 In one particular year, the number of hours of sunlight recorded each day in December at a particular weather recording location were as follows.

			1 Dec	2 Dec	3 Dec	4 Dec
			11.2 h	9.4 h	8.5 h	12.2 h
5 Dec	6 Dec	7 Dec	8 Dec	9 Dec	10 Dec	11 Dec
19.2 h	11.2 h	11.8 h	10.4 h	8.7 h	10.3 h	9.1 h
12 Dec	13 Dec	14 Dec	15 Dec	16 Dec	17 Dec	18 Dec
11.1 h	10.0 h	12.0 h	11.2 h	12.9 h	13.1 h	12.0 h
19 Dec	20 Dec	21 Dec	22 Dec	23 Dec	24 Dec	25 Dec
11.7 h	9.3 h	9.0 h	11.1 h	12.8 h	10.3 h	13.2 h
26 Dec	27 Dec	28 Dec	29 Dec	30 Dec	31 Dec	
12.1 h	7.3 h	5.2 h	9.9 h	10.1 h	11.3 h	

Calculate the mean and the median number of hours of sunlight per day for the December of this particular year at this location.

16 The scores obtained by 30 students in an exam are shown in the dot frequency diagram below. The exam was marked out of 120.

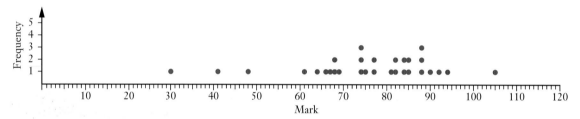

Calculate   **a**   the mean of these scores (correct to one decimal place),

          b   the median of these scores.

          c   the number of students scoring less than 60%.

          d   the percentage of students scoring greater than 75%.

17 The heights of a group of 29 students, measured to the nearest centimetre, were as shown in the stem and leaf diagram below.

Girls							Boys						
					18	6							
				5	17	9	3	0	5	2			
	8	3	5	0	16	9	6	3	2	8	5	9	
7	9	5	8	5	15	9	5	9					
			9	8	14	9							

  a   How many girls were there in the group?

  b   How tall was the shortest girl?

  c   Find the mean height for the girls (to nearest cm).

  d   Find the mean height for the boys (to nearest cm).

  e   Find the mean height for the group of 29 students (to nearest cm).

  f   Display the heights of the 29 students as a dot frequency diagram, making some distinction on your diagram between boys and girls.

18 a Calculate the mean of the 50 scores shown below.

23	42	40	47	31	42	39	31	43	26
33	50	23	40	49	30	42	40	29	37
44	48	43	26	36	43	50	32	31	44
47	31	45	37	48	32	42	41	43	23
28	47	36	26	36	45	23	49	29	48

b Rearrange the above 50 scores as grouped data using the class intervals:

$21 - 25, 26 - 30, 31 - 35, 36 - 40, 41 - 45$ and $46 - 50$.

Use the midpoints of the intervals to determine the mean for this grouped data.

19 One hundred students were asked to note the number of hours they study in the 'study week' they are given prior to an examination period. The frequency table on the right shows the results of this survey.

a Find the modal class for the number of hours spent studying in the week.

b Use the class midpoints to determine the mean for the distribution.

Number of hours (h)	Number of students
$0 \leq h < 10$	3
$10 \leq h < 20$	4
$20 \leq h < 30$	10
$30 \leq h < 40$	20
$40 \leq h < 50$	29
$50 \leq h < 60$	18
$60 \leq h < 70$	9
$70 \leq h < 80$	4
$80 \leq h < 90$	2
$90 \leq h < 100$	1

20 As part of the process of assessing the value of a block of land a real estate agent considers other blocks recently sold in the area. The agent is able to access such information from data held in his computer. For 68 recent sales the information was as follows:

Price ($C)	Midpoint of interval	Frequency
$200000 \leq C < 210000$	205000	2
$210000 \leq C < 220000$	215000	13
$220000 \leq C < 230000$	225000	10
$230000 \leq C < 240000$	235000	15
$240000 \leq C < 250000$	245000	7
$250000 \leq C < 260000$	255000	7
$260000 \leq C < 270000$	265000	4
$270000 \leq C < 280000$	275000	6
$280000 \leq C < 290000$	285000	3
$290000 \leq C < 300000$	295000	1

a In which class interval does the median price of the 68 blocks lie?

b Use the interval midpoints to calculate the mean price (to nearest $1000).

Describing a distribution of scores

The last two questions of the first exercise in this chapter, **Exercise 2A**, required you to write some sentences describing some distributions. Indeed one of those questions suggested the following useful words that you could consider using:

lowest score	highest score	tightly packed	spread out
clusters	gaps	outliers	uniform
groups	dense regions	symmetrical	

Let us now formalise what aspects we should consider when asked to describe a data distribution.

If asked to write a description of a set of scores we should comment on such things as

- the location of the scores
- how spread out they are
- the 'shape' of the distribution

and
- anything else of relevance.

- **Location:** The mean and the median of the scores give information about location.

- **Spread:** The range of the scores gives information about their spread.

- **Shape:** Symmetry, gaps, clustering, more dense/less dense regions, outliers, modality (does the data have one modal class or is it perhaps bimodal with two modal classes) all convey information about the shape of a distribution.

Note: From work of earlier years, some readers may also be familiar with *standard deviation* as a measure of spread, and with the idea that a distribution can be *skewed*. These terms are not included in the descriptions of distributions given in this section because later chapters cover these concepts. (Standard deviation is covered in chapter 3 and skewness in chapter 4.)

EXAMPLE 7

The histogram below shows the distribution of scores achieved by the students of a school in a mathematics exam.

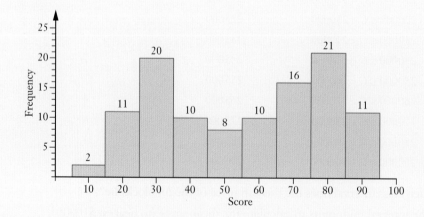

Describe the distribution of marks in this exam.

ISBN 9780170390262

One hundred and nine students from the school sat the exam.

Using the mid-point of each class interval gives an estimated mean of 55.3 and the middle ranked student, i.e. the 55th student, achieved a mark between 55 and 65.

The scores were well spread out from about 5 to 95, i.e. the range was about 90.

The scores were reasonably symmetrically spread about a mid point of about 55. The distribution of scores was bimodal peaking around 30 and again around 80.

Approximately 30% of the students scored less than 35 and approximately 30% scored more than 75.

←	Any relevant information perhaps not covered in location, spread and shape?
←	Comment about location.
←	Comment about spread.
←	Comment on shape.
←	Anything else you notice of relevance.

Exercise 2C

Describe each of the distributions shown in questions 1 to 6.

1

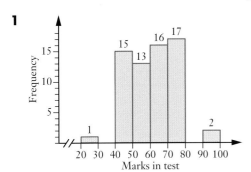

2

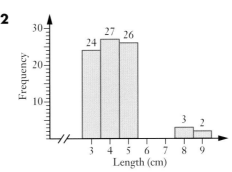

3

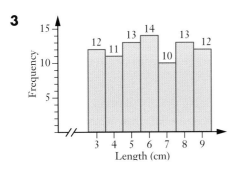

4

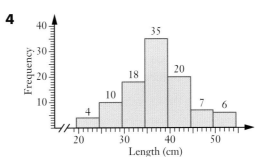

5

Score	21–25	26–30	31–35	36–40	41–45	46–50	51–55
Frequency	35	21	17	12	7	3	1

6

Score (x)	$0 \le x < 10$	$10 \le x < 20$	$20 \le x < 30$	$30 \le x < 40$	$40 \le x < 50$	$50 \le x < 60$	$60 \le x < 70$
Frequency	28	15	7	5	9	17	25

Organising, describing and interpreting data

Lizards

A survey of a particular species of lizard involved capturing about one hundred different lizards of the species, measuring their length, marking them with an identifiable tag and then releasing them back into the wild. (The tagging was to ensure that the same lizard did not feature more than once in the measuring process and also to allow for follow up studies to be carried out later.)

The lizard lengths, recorded in millimetres, are shown below (the shortest and longest lengths in the list are shown in bold):

114	140	161	110	88	153	112	163	107	151
54	113	39	115	116	106	69	165	103	104
100	110	173	93	158	109	160	60	175	112
160	155	117	162	**33**	168	152	106	156	159
90	116	104	154	161	101	164	111	107	103
167	110	158	109	115	154	42	163	158	106
105	104	47	113	163	77	158	167	**197**	119
162	156	147	160	105	163	109	108	162	111
156	63	104	104	170	131	108	115	111	153
107	122	164	116	102	162	57	160	103	115
158	105								

- Organise the data into a grouped frequency table involving what you consider to be an appropriate number of equal width intervals.

- Display the data graphically and write some sentences describing the distribution of lengths using suitable statistical vocabulary, e.g. range, outliers, clusters, modal class, mean length, etc.

- If the distribution of lengths shows any particularly notable features suggest some possible reasons why this might be the case. (The scientific team can then consider exploring such suggestions in further surveys.)

ISBN 9780170390262

Miscellaneous exercise two

This miscellaneous exercise may include questions involving the work of this chapter, the work of any previous chapters, and the ideas mentioned in the Preliminary work section at the beginning of the book.

1 Expand each of the following and simplify where possible.

   a $3(2x + 5)$ **b** $5(7x - 3)$

   c $-2(1 - 5x)$ **d** $6(2x + 1) + 5(2x - 3)$

   e $2(2x + 1) - 3(2x - 3)$ **f** $3(1 - 2x) + 2(5x + 3)$

   g $2(2x + 3) - (5x + 1) + 2x$ **h** $5(1 + 2x) - 2(3 - 2x)$

2 In a National Heart Foundation survey, 1915 women and 1863 men, all aged between 40 and 60, ticked either Yes or No when asked the question:

        In the past 2 weeks, did you walk for recreation or exercise?   *No* ☐   *Yes* ☐

   What type of variable is involved here?

3 List advantages and disadvantages of using the mean as the representative score in a set of scores. Do the same for using the median in this way and then for using the mode in this way.

4 To gain a pass a student needs to achieve a mean of at least 60% in eight tests. In the first seven tests the student achieved a mean of 54%. What percentage must the student achieve in test eight if they are to pass the course?

5 In each part of this question two dot frequency graphs are given, Graph A and Graph B. In each pair the two graphs involve sets of data with the same number of data points. However, in each case the two graphs are quite different. Write some sentences comparing datasets A and B in each case and mentioning such things as means, medians, lowest and highest scores, range, gaps, spread out, clusters, percentages of scores, outliers.

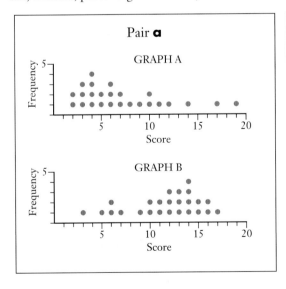

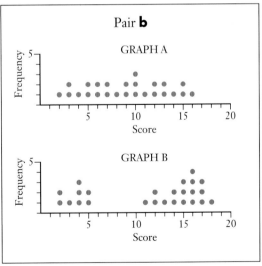

6 List six prices for secondhand cars with a mean of $28 000 but for which this mean value is not a particularly good choice to represent a price that is centrally located with regard to the six prices.

7 HEALTH SURVEY

One hundred men and one hundred women took part in a health survey. Fifty of the men and fifty of the women were aged between 20 and 25 and the remaining fifty men and fifty women were aged between 65 and 70.

One of the variables measured was the diastolic blood pressure of these two hundred people. The results of such measurements are shown below with the blood pressures given to the nearest millimetre of mercury (mmHg).

Males (aged 20 to 25)	
Blood pressure	Frequency
50 –59	1
60 – 69	10
70 – 79	21
80 – 89	14
90 – 99	3
100 – 109	1
110 – 119	0
Total	50

Males (aged 65 to 70)	
Blood pressure	Frequency
50 –59	0
60 – 69	3
70 – 79	12
80 – 89	17
90 – 99	13
100 – 109	4
110 – 119	1
Total	50

Females (aged 20 to 25)	
Blood pressure	Frequency
50 –59	4
60 – 69	22
70 – 79	19
80 – 89	5
90 – 99	0
100 – 109	0
110 – 119	0
Total	50

Females (aged 65 to 70)	
Blood pressure	Frequency
50 –59	1
60 – 69	3
70 – 79	13
80 – 89	20
90 – 99	10
100 – 109	2
110 – 119	1
Total	50

a Use the centre of each class interval to determine a mean blood pressure for each of these four groups.

b Draw frequency histograms for the two male groups.

c Draw frequency histograms for the two female groups.

d Comment on any trends suggested from parts **a**, **b** and **c**.

Shutterstock.com/Tyler Olson

Measures of
dispersion or
spread

- Standard deviation
- Use of statistical functions on a calculator
- Frequency tables
- Outliers
- Grouped data
- Central tendency and spread –
  An investigation
- Miscellaneous exercise three

Situation

The heights of the 'starting 5' players of two basketball teams are given below.

Team A		Team B	
Player 1	211 cm	Player 1	186 cm
Player 2	184 cm	Player 2	184 cm
Player 3	184 cm	Player 3	184 cm
Player 4	172 cm	Player 4	184 cm
Player 5	169 cm	Player 5	182 cm
Mean height	184 cm	Mean height	184 cm
Median height	184 cm	Median height	184 cm
Modal height	184 cm	Modal height	184 cm

Note that the two teams have the same mean as each other, the same median as each other and the same mode as each other. Can we conclude that, with regard to heights, the two teams are similar?

The situation above shows that whilst averages can be very useful in summarising data they do not tell the whole story. We also need to consider how widely the data is **spread** or **dispersed**. Thus as well as being able to summarise data using means and medians as *measures of central tendency* we also need some *measures of dispersion*. Of course we do already have one such measure of dispersion – the range of the scores. However, as was stated in the *Preliminary work* section at the beginning of this text '*Whilst the range is easy to calculate, it is determined using just two of the scores and does not take any of the other scores into account*'. Hence the range is not that useful for comparing the spread of distributions.

For example notice that the two distributions shown below, each involving 40 scores, have the same range but show very different spread patterns.

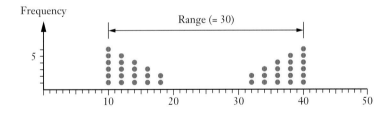

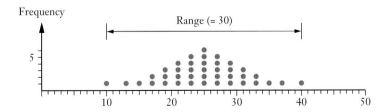

This reliance on just the lowest and highest scores makes the range of limited use and so we need to consider other measures of spread.

We will now consider two other ways of quantifying spread, namely

the mean deviation,

and   the standard deviation.

Each of these considers how much each score deviates from the mean score. In this way we can obtain measures that will tell us how concentrated the scores are about the mean value and that use each and every score in the data set in their determination.

Consider again the heights of team A from the previous page and listed again below:

Team A (mean 184 cm)		
Player	**Height**	**Deviation from the mean**
1	211 cm	211 cm – 184 cm = +27 cm
2	184 cm	184 cm – 184 cm = 0 cm
3	184 cm	184 cm – 184 cm = 0 cm
4	172 cm	172 cm – 184 cm = –12 cm
5	169 cm	169 cm – 184 cm = –15 cm

If we sum the deviations from the mean the answer is zero (as you may have expected from your understanding of the mean). Thus we cannot find the average of these deviations as they are. To avoid this problem we could:

- ignore the negative signs and find the average of the absolute values of the deviations. This technique gives the **mean deviation** of the heights.

$$\text{For team A, mean deviation of heights} \quad = \quad \frac{27+0+0+12+15}{5}$$

$$= \quad 10.8$$

Alternatively we could:

- square the deviations, find the average of these square deviations (this is called the **variance** of the scores) and then square root this variance. (This final step of finding the square root is to give a measure that has the same units as the original deviations that had been squared.)

This technique gives the **standard deviation** of the scores.

$$\text{For team A, variance of heights} \quad = \quad \frac{(27)^2+(0)^2+(0)^2+(-12)^2+(-15)^2}{5}$$

$$= \quad 219.6$$

$$\text{Thus the standard deviation} \quad = \quad \sqrt{219.6}$$

$$= \quad 14.8, \text{ correct to 1 decimal place.}$$

Note: Finding the standard deviation may seem a more complicated process than that of finding the mean deviation. However it is the standard deviation that is the more commonly used measure of dispersion in data analysis.
For this reason we will focus our attention on the standard deviation and will not pursue the concept of a mean deviation.

Standard deviation

EXAMPLE 1

Find the mean, the variance and the standard deviation of the set of scores:

$$4, 7, 10, 13, 21.$$

Solution

$$\text{Mean} = \frac{4+7+10+13+21}{5}$$

$$= 11$$

$$\text{Variance} = \frac{(4-11)^2+(7-11)^2+(10-11)^2+(13-11)^2+(21-11)^2}{5}$$

$$= 34$$

$$\text{Standard deviation} = \sqrt{34}$$

$$= 5.83, \text{ correct to two decimal places.}$$

The scores have a mean of 11, a variance of 34 and a standard deviation of 5.83 (correct to 2 decimal places).

Exercise 3A

Just as the various measures of central tendency can be obtained using the statistical capabilities of many calculators so too can some of the measures of dispersion. You will be encouraged to use your calculator to obtain such measures soon but for this exercise obtain the variance and standard deviation 'the long way', as in the above example, to gain understanding of the concepts.

1 Find the range of each of the following sets of scores.

   a   5, 7, 11, 12, 17, 19, 21, 36.

   b   104, 115, 117, 117, 118, 121, 122, 125, 125, 146.

   c   121 000, 109 000, 128 000, 90 000, 110 000, 95 000, 112 000, 107 000.

2 Find the variance of each of the following sets of scores.

   a   5, 5, 6, 8, 10, 12, 12, 13, 14, 15.

   b   7, 9, 10, 12, 15, 18, 18, 19.

3 Find the standard deviation of each of the following sets of scores.
(Give answers correct to 2 decimal places when rounding is necessary.)

   a   15, 15, 15, 15, 15, 15, 15.

   b   5, 7, 10, 11, 13, 13, 16, 21.

   c   104, 115, 117, 117, 118, 121, 122, 125, 125, 146.

   d   65, 67, 72, 83, 84.

For questions 4 to 7 state which of the two diagrams shows the set of scores with

a the greater mean   **b** the greater standard deviation.

(You should **not** need to calculate the means and standard deviations.)

4

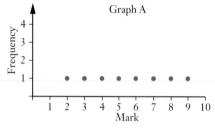
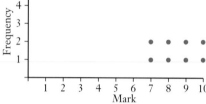

5

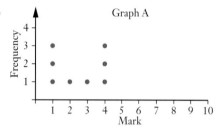
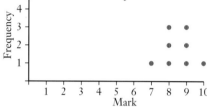

6

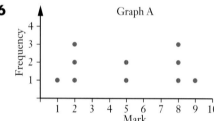
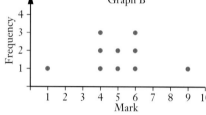

7

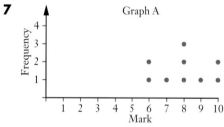
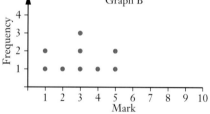

8 For each of parts **a** to **d** given below, two sets of scores are given. For each part state which set, I or II, has the greater standard deviation (you should **not** need to calculate the standard deviations).

a   Set I: 6,   6,   6,   6,   6,   6,   6   mean = 6

     Set II: 4,   5,   6,   6,   6,   7,   8   mean = 6

b   Set I: 6,   8,   9,   9,   10, 11, 11, 12, 14   mean = 10

     Set II: 6,   6,   6,   9,   10, 11, 14, 14, 14   mean = 10

c   Set I: 20, 21, 21, 22, 23, 24, 25, 25, 26   mean = 23

     Set II: 13, 15, 21, 22, 23, 24, 25, 31, 33   mean = 23

d   Set I: 1,   1,   2,   2,   10, 18, 18, 19, 19   mean = 10

     Set II: 1,   1,   9,   9,   10, 11, 11, 19, 19   mean = 10

9 Find the range, the mean and the standard deviation of the following set of scores.
(Give answers correct to two decimal places if rounding is necessary.)

$$8, 8, 9, 11, 11, 15, 16, 19, 19, 20, 20, 21, 21, 25, 32.$$

Use of statistical functions on a calculator

As was stated at the beginning of the previous exercise, just as the various measures of central tendency can be obtained using the statistical capabilities of many calculators so too can some of the measures of dispersion. Also, you should already know how to put data into a calculator and output statistical measures such as the mean and the median.

The standard deviation of a set of scores can similarly be obtained using your calculator.

We tend to use either s or σ as symbols for standard deviation. σ is a letter from the Greek alphabet and is pronounced sigma. It is a 'lower case' sigma, capital sigma is written Σ.

The diagram below shows a typical graphic calculator display for the data set:

$$3 \quad 5 \quad 2 \quad 0 \quad 2$$

1-Variable		
$\bar{x}$	= 2.4	$\longleftarrow$ The mean of the scores
Σx	= 12	$\longleftarrow$ The sum of the scores
Σx^2	= 42	$\longleftarrow$ The sum of the squares of the scores
$x\sigma n$	= 1.62480768	$\longleftarrow$ The standard deviation of the scores
$x\sigma n-1$	= 1.81659021	$\longleftarrow$ The different standard deviation – see note next page
n	= 5	$\longleftarrow$ The number of scores

Scrolling down such a display would allow further statistical information for this set of scores to be viewed.

For this set of scores: Mean = 2.4

Standard deviation = 1.625 (correct to 3 decimal places)

> **Hint**
>
> Make sure that **you** can obtain these two values from **your** calculator.

Note   • The standard deviation is a measure of spread. For most distributions very few, if any, of the scores would be more than three standard deviations from the mean, i.e. the vast majority of the scores (and probably all of them) would lie between

$$\bar{x} - 3\sigma \text{ and } \bar{x} + 3\sigma$$

Indeed we would frequently find that about two thirds of the scores would lie within 1 standard deviation of the mean, i.e. between $\bar{x} - \sigma$ and $\bar{x} + \sigma$.

Note • The calculator display shown has two different standard deviations,

$$\sigma_n \text{ and } \sigma_{n-1}$$

σ_n is the standard deviation of the five scores.

σ_{n-1}, sometimes shown as s_x, gives an answer a little bigger than σ_n by dividing the sum of the squared deviations by $(n-1)$ rather than n. This would be used if the five scores were a sample taken from a larger population and we wanted to use the deviation of the sample to estimate the standard deviation of the whole population, as is often the case in real life. Division by $(n-1)$ rather than n compensates for the fact that there is usually less variation in a *small* sample than there is in the entire population. (This is known as Bessel's correction.) If we make the sample large then n will be large and there will be little difference between σ_n and σ_{n-1}. This book makes the distinction between σ_n and σ_{n-1} and if a question asks for the standard deviation of a set of scores to be determined σ_n is given, unless the question *specifically* states that the task is to use the small sample data to estimate the standard deviation of the population of which the sample is a part. However in some states of Australia this distinction may not be part of the course and instead you might simply be expected to give σ_{n-1}, or s_x, whenever standard deviation is requested. Hence it is important that you make sure you know which standard deviation, σ_n or σ_{n-1}, you are expected to give in examinations when a question simply asks you to determine a standard deviation. To assist readers in states that only require σ_{n-1} to be determined answers in the back of this text will tend to give both values for those cases where, to the given accuracy, the values differ.

EXAMPLE 2

Find the mean and standard deviation for the following set of scores.

$$9, 10, 11, 13, 19, 20, 21, 21, 22, 24, 25, 31, 31, 32, 44.$$

a How many of the scores are such that

$$\bar{x} - 1 \text{ standard deviation} < \text{score} < \bar{x} + 1 \text{ standard deviation?}$$

b How many of the scores are such that

$$\bar{x} - 2 \text{ standard deviations} < \text{score} < \bar{x} + 2 \text{ standard deviations?}$$

Solution

Using a calculator $\bar{x} = 22.2$, standard deviation $(\sigma_n) = 9.27$ (2 decimal places)

a $\bar{x}$ − standard deviation ≈ 12.93 and $\bar{x}$ + standard deviation ≈ 31.47.

10 of the 15 scores lie between $\bar{x} - \sigma$ anwd $\bar{x} + \sigma$.

b $\bar{x}$ − 2 standard deviations ≈ 3.66 and $\bar{x}$ + 2 standard deviations ≈ 40.74.

14 of the 15 scores lie between $\bar{x} - 2\sigma$ and $\bar{x} + 2\sigma$.

ISBN 9780170390262

Standard deviation

Exercise 3B

Find the mean and the standard deviation for each of the sets of scores given in numbers 1 to 5.
(Give answers correct to one decimal place when rounding unless stated otherwise.)

1 10, 11, 12, 13, 14, 15.

2 15, 26, 47, 16, 33, 49, 8, 11, 41, 26, 19, 14.

3 31, 29, 33, 32, 34, 29, 30, 30.

4 6.6, 6.2, 7.3, 8.1, 6.8, 7.0, 6.9, 7.1, 6.9, 7.0. (Answers correct to 2 decimal places)

5 30, 29, 34, 27, 26, 25, 26, 38, 38, 23, 39, 35, 26, 27, 29, 32, 29, 31, 32, 30, 31, 27, 29, 32, 30, 32, 31, 28, 32, 30, 29, 30.

6 Three groups of 10 students do a spelling test marked out of 10. The scores achieved by each group are shown in the dot frequency graphs below.

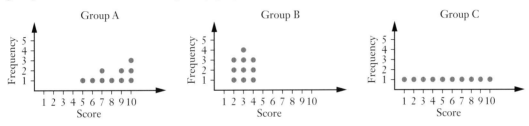

a Without calculating values but just by looking at the graphs state which of the three groups have scores with

    i the greatest standard deviation   **ii** the smallest standard deviation

    iii the greatest mean   **iv** the smallest mean.

b Calculate the mean and standard deviation for each group.

7 In a particular sporting competition contestants are awarded a score by each of eight judges. The eight scores for one competitor were:

    5.9   5.9   5.7   6.0   5.1   5.8   5.8   5.9

a Find the mean and standard deviation of these scores. (To 2 decimal places.)

b If the highest score and the lowest score are discarded find the mean and standard deviation of the remaining scores. (To 2 decimal places.)

8 An engineering company makes a particular component that theoretically is to be of length 63 cm. Quality control imposes certain restrictions that any randomly selected sample of ten of these components must satisfy. One such random sample has lengths (in cm):

    63.0,   62.9,   63.0,   63.2,   63.0,   63.1,   63.0,   62.9,   63.1,   63.1.

a Does this sample satisfy the restriction: 62.95 cm < mean < 63.05 cm?

b Does this sample satisfy the restriction: standard deviation < 0.1 cm?

9 The 25 students in a year 8 class have the following heights (nearest cm).

181 145 162 158 165 150 164 155 173 160 164 154 161
152 167 169 148 163 175 153 166 153 166 147 160

a Find the mean and standard deviation of these 25 heights (to 2 decimal places).

b These 25 heights are to be used to estimate the standard deviation of the 231 year 8 students in this school. What would this estimated standard deviation be (to 2 decimal places)?

10 An entomologist catches an adult moth that he is sure is one belonging to a particular species but he is not sure whether it is type A of the species or type B. Some of the smaller examples of type B moths can easily be mistaken for a type A moth and positive examination then requires analysis of body tissue. The entomologist refers to a reference book which states that in an extensive survey involving thousands of these adult moths it was found that the body length of the two types were such that:

For the type A moths surveyed: Mean 15 mm, Standard deviation 1mm.
For the type B moths surveyed: Mean 22 mm, Standard deviation 3mm.

The entomologist measures the body length of 'his' moth as 18 mm.

Decide which it is more likely to be, a large type A or a small type B, and explain your choice.

11 A scientist collects 40 butterflies of a particular species and measures the lengths of the wing span of each one. The lengths, to the nearest millimetre, are shown in the dot frequency diagram below.

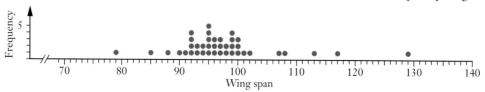

a Calculate the mean and standard deviation (st. dev^n) for this set of lengths, giving the standard deviation correct to 3 decimal places.

b What percentage of the 40 lengths lie within one standard deviation of the mean?
i.e. What percentage of the lengths are such that:
$$(mean - 1 \text{ st. } dev^n) < length < (mean + 1 \text{ st. } dev^n)?$$

c What percentage of the 40 lengths lie within two st. dev^{ns} of the mean?

d What percentage of the 40 lengths lie within three st. dev^{ns} of the mean?

e If the scientist wanted to use these 40 lengths to estimate the standard deviation of the entire butterfly population of this species what would this estimated standard deviation be (to 2 decimal places)?

12 A scientific experiment involved students determining the temperature at which a particular chemical reaction took place. Ten groups carried out the experiment and obtained the following answers:

156°C, 163°C, 154°C, 158°C, 159°C, 161°C, 121°C, 163°C, 159°C, 159°C.

One of the groups discovered that they had made a number of errors in carrying out the experiment and in calculating the answer.

Determine the mean and standard deviation of the results if

a the answer likely to be from the group making errors **is** included,

b the answer likely to be from the group making errors is **not** included.

ISBN 9780170390262

13 Twenty-four of the twenty-five students in a class sat a maths test that was marked out of 40. The marks obtained were as shown below:

22 25 21 18 25 32 30 40 28 16 31 21
24 14 25 34 37 27 18 27 39 35 28 35

The twenty-fifth student was absent for the test due to hospitalisation and so the teacher had to estimate a mark for this student in this test. Noticing that on previous tests this student usually performed above the class mean the teacher awards an estimated mark of $(\bar{x} + 0.6s)$ where $\bar{x}$ and s are respectively the mean and standard deviation of the 24 marks. To the nearest 0.5 of a mark what was the student's estimated mark?

14 The maximum and minimum temperatures, as recorded at Perth airport, for each day of December in a particular year were as follows:

Day	1st	2nd	3rd	4th	5th	6th	7th	8th	9th	10th	
Max (°C)	27.3	27.9	29.4	29.5	30.2	31.4	33.9	34.4	21.9	23.3	
Min (°C)	13.9	14.0	15.7	14.6	14.6	15.0	17.0	24.9	16.1	12.1	
Day	11th	12th	13th	14th	15th	16th	17th	18th	19th	20th	
Max (°C)	27.0	27.2	24.7	26.2	34.5	39.2	41.2	36.2	28.7	24.1	
Min (°C)	13.0	13.0	12.5	12.9	13.6	19.8	20.8	15.2	18.6	15.0	
Day	21st	22nd	23rd	24th	25th	26th	27th	28th	29th	30th	31st
Max (°C)	26.0	29.1	29.2	30.0	33.0	37.4	38.6	24.7	28.7	36.0	31.9
Min (°C)	10.5	13.1	13.6	15.4	16.0	16.7	23.3	19.4	14.0	16.6	19.8

[Source of data: Bureau of Meteorology.]

a Determine the mean, range and standard deviation of the daily maximum temperatures featured in the above table, giving your answers correct to one decimal place.

b Data collected over a period of more than 100 years, prior to the year featured above, gave Perth's mean maximum daily temperature for December as 27.4°C. Compare your mean daily maximum temperature for the above table with this long-term mean.

c Determine the mean, range and standard deviation of the daily minimum temperatures featuring in the above table, giving your answers correct to one decimal place.

d Data collected over a period of more than 100 years, prior to the year featured in the table, gave Perth's mean minimum daily temperature for December as 16.3°C. Compare your mean daily minimum temperature for the above table with this long-term mean.

S. Forster / Alamy Stock Photo

15 The marks achieved in an exam sat by 100 candidates are shown below.

53	58	45	61	89	55	60	49	62	26	65	92	51	59	40
56	21	61	65	56	80	40	69	54	83	59	83	47	77	36
46	58	62	52	69	97	66	64	75	14	77	62	56	58	81
64	36	69	64	66	72	47	80	50	70	56	43	68	52	28
40	69	52	78	32	78	67	56	83	62	43	67	64	56	85
62	49	59	89	62	66	42	62	53	47	85	74	86	79	75
70	75	53	72	23	70	77	74	80	71					

a Calculate the mean and standard deviation for this set of marks (correct to 1 decimal place).

b Display the data as a dot frequency graph.

c Show on your graph the 'grade borderlines' and state the number of candidates awarded each grade given that grades were awarded as follows:

A: exam mark $\geq$ (mean + 1.25 × st. dev^{n}.)

B: (mean + 0.5 × st. dev^{n}.) $\leq$ exam mark < (mean + 1.25 × st. dev^{n}.)

C: (mean – 0.5 × st. dev^{n}.) $\leq$ exam mark < (mean + 0.5 × st. dev^{n}.)

D: (mean – 1.5 × st. dev^{n}.) $\leq$ exam mark < (mean – 0.5 × st. dev^{n}.)

F: exam mark < (mean – 1.5 × st. dev^{n}.)

16 A company manufactures components for aircraft engines. The quality control for one particular component involves 25 of the components being randomly selected and measured from every batch of 500. If **any** one of the conditions stated below is found to apply to the sample then production is halted, each of the other 475 components in the batch is checked and the machine is reset.

Sample reject, condition 1:	mean < 64.8 cm
Sample reject, condition 2:	mean > 65.2 cm
Sample reject, condition 3:	standard deviation > 0.15 cm
Sample reject, condition 4:	any one component > 65.3 cm
Sample reject, condition 5:	any one component < 64.7 cm

For each of the following samples of 25, determine whether the sample passes these checks and, for any sample that does not pass, state the reason for it not passing. (All measurements are in centimetres.)

Sample A				
65.0	65.0	64.7	64.9	65.1
65.1	65.0	65.0	65.1	65.0
65.0	64.8	64.9	65.0	64.8
64.8	65.0	65.0	65.1	65.0
65.0	64.9	65.1	65.0	65.2

Sample B				
65.0	65.0	65.0	65.0	65.0
65.1	65.1	65.1	64.9	65.0
65.0	65.0	64.8	64.6	64.9
64.9	65.1	65.0	64.9	65.0
65.0	65.0	64.8	65.0	65.0

Sample C				
64.9	64.9	64.8	65.1	65.0
64.9	65.0	65.1	65.0	65.0
65.0	65.1	65.0	65.1	65.0
65.1	65.1	64.9	65.0	64.8
65.1	65.0	65.1	65.0	65.0

Sample D				
64.8	65.2	65.0	65.0	65.2
65.0	64.8	65.1	65.2	64.8
65.0	65.1	65.1	64.8	65.2
64.7	64.8	64.8	65.2	65.0
64.9	65.1	65.0	65.2	64.8

ISBN 9780170390262

Frequency tables

Remember that if data is given in the form of a frequency table it can be entered into many calculators in this frequency form.

For example, given the following table

Score	10	11	12	13	14	15	16	17	18	19	20
Frequency	1	0	1	2	4	9	10	11	8	3	1

the 50 scores can be keyed into a calculator in this frequency form. We do not need to key in the 50 scores separately.

	List 1	List 2	List 3	List 4
1	10	1		
2	11	0		
3	12	1		
4	13	2		
5	14	4		
6	15	9		

```
1-Variable
x̄          = 16.14
Σx         = 807
Σx²        = 13205
xσn        = 1.897472
xσn–1      = 1.91673617
n          = 50
```

Outliers

Any data values that are unusually far away from the others are known as **outliers**. These extreme values can have a big effect on the standard deviation.

Suppose for example that the frequency table shown above were also to include one score of 51.

Score	10	11	12	13	14	15	16	17	18	19	20	51
Frequency	1	0	1	2	4	9	10	11	8	3	1	1

Now the standard deviation is 5.2 (to 1 decimal place), as shown on the right, compared to the previous value of 1.9 without the outlier.

```
1-Variable
x̄          = 16.8235294
Σx         = 858
Σx²        = 15806
xσn        = 5.18559801
xσn–1      = 5.23719727
n          = 51
```

For this reason outliers need careful consideration. Perhaps in this case it was simply an error involving a score of 15 being written as 51.

Sometimes we might determine the standard deviation without the outlier included and then comment on the presence of the outlier.

Calculating and interpreting summary statistics

Grouped data

Just as we use the midpoint of each interval to estimate the mean for grouped data, we do the same thing to obtain an estimate for the standard deviation.

Score	20 – 29	30 – 39	40 – 49	50 – 59	60 – 69	70 – 79	80 – 89
Frequency	3	11	18	28	24	10	6

$$n = 100, \quad \bar{x} = 55.8, \quad \sigma_n = 14.3 \text{ (1 decimal place).}$$

L1	L2	L3
24.5	3	
34.5	11	
44.5	18	
54.5	28	
64.5	24	
74.5	10	
84.5	6	

```
1-Var Stats
x̄        = 55.8
Σx       = 5580
Σx²      = 331895
Sx       = 14.4008277
σx       = 14.32864264
n        = 100
```

Remember: Many measured quantities are naturally grouped by rounding and the data is often displayed as a histogram.

Confirm that for the histogram shown on the right:

Mean = 50.88,
Standard deviation = 1.5 (1 decimal place).

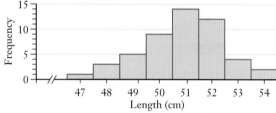

Exercise 3C

Find the mean and standard deviation of each of the distributions shown in questions 1 to 5. (Give answers correct to one decimal place.)

1

Score	0	1	2	3	4	5
Frequency	3	7	15	24	19	12

2

Score	15	20	25	30	35	40	45	50
Frequency	1	3	4	8	8	4	3	1

3

Score	15	20	25	30	35	40	45	50
Frequency	8	4	3	1	1	3	4	8

4

Score	1	2	3	4	5	6	7	8	9	10
Frequency	1	0	4	6	10	4	2	1	1	1

5

Score	10	20	30	40	50	60	70	80	90	100
Frequency	1	2	4	3	5	7	15	19	12	7

ISBN 9780170390262

Use the midpoint of each class interval to determine the mean and standard deviation of the following distributions shown in questions 6 to 10. (Give answers correct to one decimal place if rounding is necessary.)

6

Score	Frequency
20 → 24	7
25 → 29	12
30 → 34	18
35 → 39	20
40 → 44	24
45 → 49	13
50 → 54	6

7

Score (x)	Frequency
$0 \leq x < 10$	17
$10 \leq x < 20$	13
$20 \leq x < 30$	9
$30 \leq x < 40$	7
$40 \leq x < 50$	4

8

Score	Frequency
0 → 9	3
10 → 19	8
20 → 29	15
30 → 39	24
40 → 49	34
50 → 59	16

9

Score (x)	Frequency
$0 \leq x < 20$	1
$20 \leq x < 40$	3
$40 \leq x < 60$	10
$60 \leq x < 80$	9
$80 \leq x < 100$	7

10

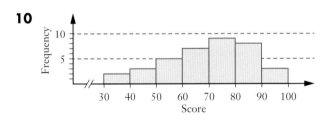

11 A golf club organises a 'club members championship' each year for the top 35 ranked players in the club. In a particular year the scores achieved by these players in the championship round were as shown below.

Score	67	69	71	72	73	74	75	76	78	82	85	91
Number of players	1	1	2	4	7	5	3	5	3	2	1	1

a How many standard deviations from the mean was the best (i.e. lowest) score? (Answer correct to one decimal place.)

b How many standard deviations from the mean was the worst (i.e. highest) score? (Answer correct to one decimal place.)

12 A company wishes to test a coating it is developing for seeds. The coating is designed to provide the seeds with essential nutrients and to stimulate germination and growth. The company arranges 200 trays each containing identical potting mix. In each of 100 of these trays 50 coated seeds are planted and in each of the other 100 trays 50 uncoated seeds are planted, all the seeds being of the same quality and type. After a certain number of weeks the company counts the number of successful germinations in each tray – a 'success' being a healthy seedling at least 6 cm in height. The results were as follows:

Uncoated Seeds		Coated Seeds	
No. of successes in tray.	**No. of trays.**	**No. of successes in tray.**	**No. of trays.**
21 – 25	7	21 – 25	1
26 – 30	21	26 – 30	2
31 – 35	28	31 – 35	10
36 – 40	23	36 – 40	17
41 – 45	15	41 – 45	42
46 – 50	6	46 – 50	28

Calculate the mean and standard deviation for each set of 100 trays and comment on your results.

13 The time that thirty patients had to wait beyond their allotted appointment time at a particular health centre was noted. The results are shown tabulated below.

Time (t mins)	$0 \leq t < 10$	$10 \leq t < 20$	$20 \leq t < 30$	$30 \leq t < 40$	$40 \leq t < 50$	$50 \leq t < 60$
No. of patients	8	15	4	2	0	1

Find the mean and standard deviation of this distribution of waiting times both with and without the outlier included. (Round to one decimal place if necessary.)

14 One hundred primary schools are surveyed regarding the number of students on the roll of each school. The information collected is shown tabulated below:

No. of students	1 to 50	51 to 100	101 to 150	151 to 200	201 to 250
No. of schools	5	5	10	9	18

No. of students	251 to 300	301 to 350	351 to 400	401 to 450	451 to 500
No. of schools	22	15	6	7	2

No. of students	501 to 550	551 to 600	601 to 650	651 to 700	701 to 750
No. of schools	0	0	0	0	1

a By taking the centre of each interval as the number of students in each school in that interval, determine estimates for the mean number of students per school and the standard deviation of the distribution.

b Determine the mean and standard deviation of the data once the outlier in the data is removed.

ISBN 9780170390262

Central tendency and spread – An investigation

Forty students sat a maths test marked out of 50. The scores they achieved are shown in the dot frequency graph below.

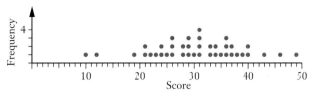

Find the mean (to 2 decimal places) and standard deviation (to 1 decimal place) of the forty scores.

The teacher wanted these scores as percentages so she multiplied each score by 2.
The dot frequency graph of the percentage scores is given below.

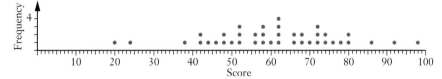

Find the mean and standard deviation of these forty percentage scores.

A second test, given to the same forty students, was in two parts. Part A consisted of 10 mental questions and was marked out of 10. Part B was more involved and was marked out of 90.
The dot frequency for the scores achieved in part B is shown below.

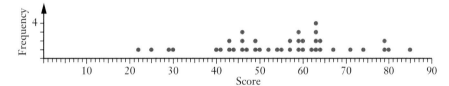

Find the mean and standard deviation of these forty scores.

Part A proved to be very straightforward and all of the students scored 10 out of 10!
The dot frequency for the combined score out of 100 is given below.

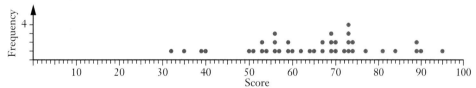

Find the mean and standard deviation of these forty scores.

Suggest how the mean and the standard deviation of a set of scores are affected
if   • the scores have the same number added to each of them (or subtracted from each them),
and/or • each of the scores is multiplied by (or divided by) the same number.

Test your suggestions (conjectures) by putting a set of scores into your graphic calculator, creating a new set of scores by adding a number to each of the scores in the first set, and/or by multiplying each set of scores in the first set by a number, and then comparing the mean and standard deviation of the two sets of scores.

Miscellaneous exercise three

This miscellaneous exercise may include questions involving the work of this chapter, the work of any previous chapters, and the ideas mentioned in the Preliminary work section at the beginning of the book.

1 Write the following numbers in order of size, smallest first.

0.201   0.12   0.102   0.21   0.1   0.021   0.012   0.2

2 Without the assistance of a calculator, write the following numbers in order of size, smallest first.

$\dfrac{1}{2}$ $\dfrac{1}{5}$ $\dfrac{1}{3}$ $\dfrac{2}{3}$ $\dfrac{3}{4}$ $\dfrac{7}{10}$ $\dfrac{1}{100}$ $\dfrac{3}{5}$

3 Evaluate each of the following.

a $3 + 2 \times 4$ **b** $5 \times 4 + 6$ **c** $3 + 5^2$

d $32 \div 4 \div 2$ **e** $32 \div (4 \div 2)$ **f** $(5 + 3)^2 - 8 \times 3$

4 State the coordinates of each of the points A to J shown on the right. (All coordinates involve integer values only.)

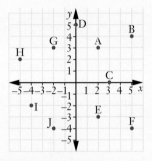

5 The statistics regarding commercial orchard fruit in Western Australia for a particular year are shown in the following table:

Fruit	Number of trees (1000s)	Production (tonnes)	Gross value of production ($1000s)
Apples	778	37 418	19 497
Pears	175	8 399	4 886
Lemons & Limes	18	1 125	738
Mandarins	55	1 315	1 830
Oranges	183	5 304	1 830
Nectarines	147	2 333	2 333
Peaches	126	2 507	4 070
Plums & Prunes	190	3 494	4 392

[Source of data: Australian Bureau of Statistics.]

a   What type of variable does the first column of the table involve?

For this particular year:

b   how many commercial apple trees were in Western Australia?

c   how many tonnes of peaches were produced commercially?

d   find the gross value per tonne for   **i**   oranges?   **ii**   nectarines?

e   on average, how many kilograms of apples did each commercial apple tree yield?

6 List one advantage and one disadvantage of using the range as an indicator of the spread or variability in a set of scores.

7 The dot frequency graph on the right shows the scores achieved by 19 students in a spelling test. Find the mean, median, mode and standard deviation of the distribution. (Round to two decimal places if rounding is necessary.)

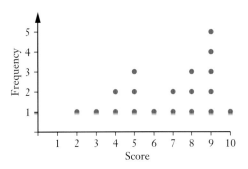

8 Ten scores have a mean of 85.4. However it was later found that one of the scores had been recorded incorrectly as 75 when it should have been 57. If this error is corrected what is the new mean of the ten scores?

9 Four maternity hospitals, A, B, C and D report to the regional health authority and state the number of live births that occurred in a particular month and the mean birth weight of the babies. The data was as follows:

Hospital A: 84 live births, mean weight 3.025 kg.
Hospital B: 27 live births, mean weight 3.140 kg.
Hospital C: 53 live births, mean weight 2.935 kg.
Hospital D: 17 live births, mean weight 2.855 kg.

Calculate the mean birth weight for all the live births from these maternity hospitals in the month that the above data applies to.

10 For a period of time, a car salesman asked customers trading in their old vehicle for a new one, how many vehicles they had owned in their lives prior to the purchase of their new one.

The responses led to the following graph:

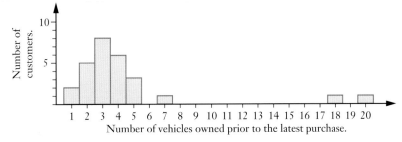

a According to the graph none of the people involved in the survey said they had owned no vehicles prior to the latest purchase. Why is this?

b The salesman concluded that, on average, people purchasing a new vehicle from him had owned approximately 4.4 vehicles before this latest purchase. Comment on this conclusion.

11 (Pythagoras)

A vertical mast is 80 metres tall and is to be held in place by a number of wires. These wires are each one of three different lengths and are classified as short, medium or long.

Each short wire is to have one end attached to a point one quarter of the way up the mast and the other to a point on the ground, level with the base of the mast and thirty metres from it.

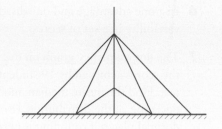

Each medium wire is to have one end attached to a point twenty metres from the top of the mast and the other to a point on the ground, level with the base of the mast and thirty metres from it.

Each long wire is to have one end attached to a point twenty metres from the top of the mast and the other to a point on the ground, level with the base of the mast and sixty metres from it.

The wires are to be made a little longer than is required, brought to the site, and then suitably adjusted. Each wire is to be made to 'the accurate length plus 50 cm then round up to the next 10 cm'.

Find the length that each classification of wire should be made to.

12 The assessment of her college course involves Suzanne sitting five exams, one in each of the units A, B, C, D and E. The twenty-five students following this course all took the five exams and their results in the exam for unit A were as follows:

55	50	54	49	14	53*	50	37	37	48
40	71	20	57	61	55	9	46	30	44
50	43	48	34	43					

*Suzanne's result.

a Determine the mean and standard deviation for these scores.

The mean and standard deviation for the marks obtained by these students in the other four exams are shown below, together with Suzanne's score.

Unit B:	Mean	67	Standard deviation	12	Suzanne's score	61
Unit C:	Mean	37	Standard deviation	8	Suzanne's score	49
Unit D:	Mean	83	Standard deviation	5	Suzanne's score	80
Unit E:	Mean	72	Standard deviation	10	Suzanne's score	79

b In order to compare her marks in the five exams Suzanne decides to standardise the marks by expressing each mark in terms of the number of standard deviations the mark is above or below the mean. For example in a course having a mean of 55 and a standard deviation of 12 then a score of 67 ($= 55 + 1(12)$) becomes 1, a score of 79 ($= 55 + 2(12)$) becomes 2, a score of 43 ($= 55 - 1(12)$) becomes -1 etc.

List the units in order from the one that Suzanne achieved her highest standardised score to the one with her lowest standardised score and state the standardised score for each unit.

ISBN 9780170390262

4.

Boxplots, histograms and more about describing distributions

- Box and whisker diagrams (boxplots)

- Boxplot or histogram?

- More about the shape of a distribution – skewness

- Miscellaneous exercise four

Box and whisker diagrams (boxplots)

A simple diagram that shows the way a set of scores is distributed is a **box and whisker diagram** or **box plot**. This type of diagram does *not* show all the individual scores (unlike a dot frequency diagram which does show all of the scores) but instead it concentrates our attention on specific features of the data.

Just as the median divides the distribution into two halves then so the **quartiles** divide the distribution into four quarters. Box and whisker diagrams show the locations of:

the **lowest score**,

the **highest score**,

the **median**,

the **lower quartile**

and   the **upper quartile**.

Using this **five-number summary**, boxplots give a visual impression of the location of the data and how widely spread it is. The range (highest score minus lowest score) and the **interquartile range** (upper quartile minus lower quartile) can be determined.

For example, consider the set of scores

12,   6,   10,   19,   9,   12,   4,   14,   8,   16,   6.

Listing the scores in order allows the quartiles and interquartile range to be determined:

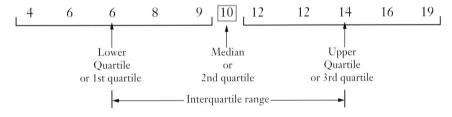

A box plot can then be drawn with the 'box' extending from the lower quartile to the upper quartile with a line in the box indicating the median. 'Whiskers' then extend from the lower quartile to the lowest score and from the upper quartile to the highest score.

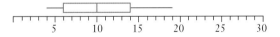

Box-and-whisker plots

Interquartile range

Boxplots 1

Boxplots 2

Five-number summaries

Statistical calculations

Statistics review

EXAMPLE 1

Draw box and whisker plots for each of the following sets of scores.

a 12, 22, 22, 23, 27, 14, 27, 23, 21, 30, 26, 17, 23, 17.

b 7, 11, 11, 11, 8, 17, 10, 12, 10, 14, 9, 15, 9.

c 21, 18, 28, 30, 23, 17, 30, 27, 28, 19, 29, 20.

Solution

a Order the scores and find the median, lower quartile and upper quartile:

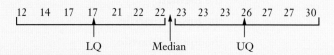

Hence draw the box plot:

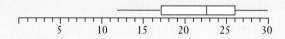

b Order the scores and find the median, lower quartile and upper quartile:

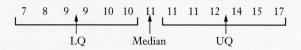

Hence draw the box plot:

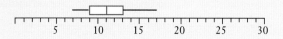

c Order the scores and find the median, lower quartile and upper quartile:

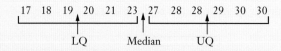

Hence draw the box plot:

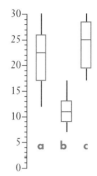

Note • The box and whisker diagrams shown in the previous example have all been drawn horizontally but they may also be drawn vertically, as shown on the right.

• Some graphic calculators can display data as box plots. The display below left shows the data of example 1 (remaining data could be viewed by scrolling down) and the display below right shows the box plots.

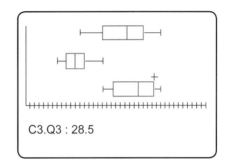

n	C1	C2	C3	C4
1	12	7	21	
2	22	11	18	
3	22	11	28	
4	23	11	30	
5	27	8	23	
6	14	17	17	
			12	

C3.Q3 : 28.5

• A possible refinement of this type of diagram is to indicate scores that may be considered as unusually high or unusually low compared to the others, as separate points on the diagram (i.e. show *outliers* as separate points.) The whiskers are drawn to include all scores that are within 1.5 times the interquartile range of the nearest quartile. Any scores outside that are considered outliers and are marked separately. The diagram below shows an example of this. The interquartile range is 8. The whiskers only extend to include marks that are no more than 12 marks (= 1.5 × 8) from the nearest quartile. Any scores beyond this are marked with a ×.

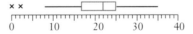

Most of the questions in this book will involve the simpler boxplots where the whiskers are drawn to the lowest and highest scores. However the idea of using 'more than 1.5 × interquartile range beyond the upper and lower quartiles' as the criteria for identifying possible outliers should be remembered.

Exercise 4A

For each of the box and whisker diagrams shown in numbers 1 to 4 state:

a the median
b the lower quartile
c the upper quartile
d the lowest score
e the highest score
f the interquartile range

1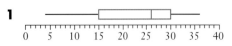

2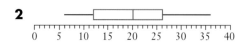

3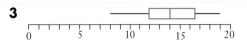

4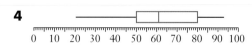

5 Four year 10 maths classes, A, B, C and D, take the same test, marked out of 100. The diagram on the right shows box plots for the results.

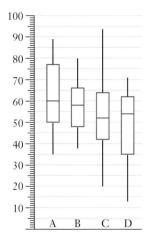

    a In which class is the student who scored the highest mark?
    b In which class is the student who scored the lowest mark?
    c Which class had the highest median?
    d Which class had the lowest median?
    e Which class had the smallest interquartile range?
    f Which class had the greatest range of marks?
    g Which class had the smallest range of marks?

Draw box plots for each of the following data sets.

6 5, 6, 11, 12, 12, 15, 16, 18, 18, 19, 20, 22, 25, 29, 31.

7 11, 14, 7, 16, 16, 5, 14, 14, 24, 7, 12, 15, 14, 9.

8 7, 10, 17, 23, 9, 12, 20, 2, 15, 5, 10, 12, 1.

9 1, 14, 11, 25, 16, 14, 1, 1, 7, 18, 20, 5.

10 The box plots on the right are for scores achieved by three classes, in the same test.

Comment on each of the following statements.

    a Class III had more scores below the median than above it.
    b The class I marks were more spread out than the class II marks.
    c The class III marks were more spread out than the class I marks.
    d The class I marks and the class II marks were similarly distributed.
    e Based on this test the top student in class III would be the twenty fifth student if they moved to class I.
    f Class III had lots of students who scored a lower mark than the lowest mark from the other two classes.

ISBN 9780170390262

Boxplot or histogram?

The lengths of the beaks of sixty male birds of a particular species were measured and the lengths, recorded to the nearest centimetre, were as follows:

iStock.com/Jens_Lambert_Photography

8	7	10	13	11	9	9	7	10	12	8	11	7	11	8
11	13	9	14	10	7	6	10	8	11	10	8	9	10	7
9	12	11	8	6	8	10	12	9	7	10	8	12	6	9
7	10	14	9	8	10	13	8	10	12	11	11	13	9	10

Below left shows the data displayed as a histogram and below right it is displayed as a boxplot.

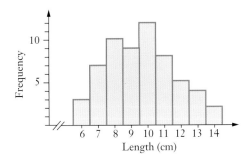
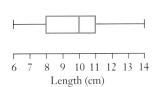

Question: Which is the better form of display?

Answer: Well, they are each useful in their own way and each allows us to visualise how the data is distributed.

Box plots can be drawn quickly, allow a five number summary consisting of lowest score, lower quartile, median, upper quartile and highest score to be readily obtained and the range and the interquartile range to be determined. Their compact nature and ease of production allows several boxplots to be drawn in close proximity thus allowing distributions to be compared easily.

Histograms convey the overall 'shape' of a distribution allowing aspects such as symmetry, grouping, gaps, modes etc to be noticed. They allow the mean and the standard deviation of the data to be determined, or at least estimated if grouped data is involved. However certain features can be hidden if we choose too few or two many class intervals.

Hence which is 'better' depends upon how much information we are wanting to show, whether we want a quickly produced visual summary of the data or a more detailed picture.

Thus box plots and histograms are both useful methods of data display, each enabling us, in their own way, to build up a picture of how a set of scores are distributed. They complement each other. Sometimes both forms of display may be given for the same set of data, as was the case above. Each form of display provides information about three key aspects of a data set:

- its **location**,   (Where is it?)
- its **dispersion**,   (How spread out is it?)
- its **shape**.   (What does it look like?)

The three aspects of **location**, **spread** (dispersion) and **shape**, together with any other information we may notice about a distribution, were the aspects of a distribution that we were encouraged to consider when describing a distribution in chapter two. The work of chapter three now allows us to include mention of standard deviation when considering spread and the following section covers **skewness**, an aspect we can consider when describing the shape of a distribution.

More about the shape of a distribution – skewness

Consider the histogram shown on the right.

Using the centre of each class interval to determine the mean and standard deviation of the distribution gives:

Mean = 115

(As we would expect considering the symmetry of the histogram.)

St. dev^n.(σ_n) = 16.04 (2 decimal places)

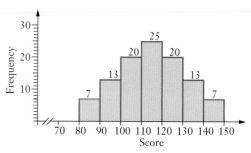

Suppose we now take the scores to the left of the central column, 40 scores in this case, and spread them further out on this left side, as shown on the second histogram.

The median score would still lie in the central column, and would therefore be unchanged, but consider what will have happened to the mean and the standard deviation?

We now have more extreme scores to the left of the centre score and these scores will drag the mean left, and increase the standard deviation.

Mean = 112.71

St. dev^n.(σ_n) = 19.58 (2 decimal places)

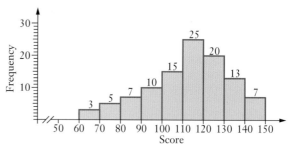

Spreading the left half of the distribution further left drags the mean further left and increases the standard deviation yet more.

Mean = 111.38

St. dev^n.(σ_n) = 22.39 (2 decimal places)

Whilst the first histogram was symmetrical we say that the second and third histograms are **skewed to the left**, also referred to as being **negatively skewed**.

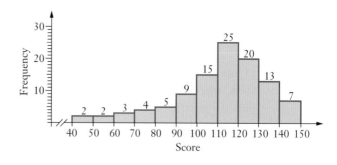

ISBN 9780170390262

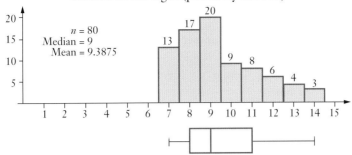

Skewed to the right (positively skewed)

If a distribution is skewed to the right, i.e. positively skewed, the longer 'tail' will be in the positive direction. The mean will usually be to the right of the median, i.e. for most positively skewed distributions we would expect

mean > median

because the 'tail' of high scores to the right will tend to drag the mean right.

The box plot will tend to be longer to the right of the median than it is to the left.

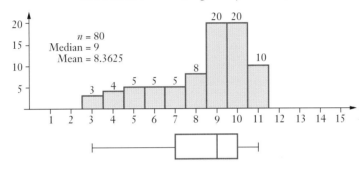

Skewed to the left (negatively skewed)

If a distribution is skewed to the left, i.e. negatively skewed, the longer 'tail' will be in the negative direction. The mean will usually be to the left of the median, i.e. for most negatively skewed distributions we would expect

mean < median

because the 'tail' of low scores will tend to drag the mean to the left.

The box plot will tend to be longer to the left of the median than it is to the right.

Note: The explanation of skewness given here is somewhat simplistic and does not, for example, consider what skewness might mean for a multimodal distribution, nor does it attempt to 'quantify' skewness. However the explanation given here is sufficient for a basic understanding of the idea.

EXAMPLE 2

Following a traffic warning that due to a number of accidents and roadworks long delays were likely to occur for people making their way home, the workers of one company decided that they would each record how long it took them to get home that evening. The histogram below shows the distribution of recorded times.

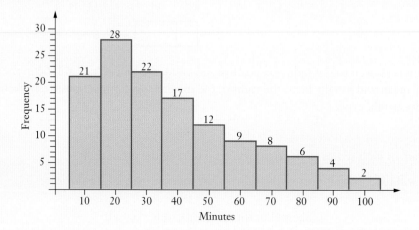

Describe the distribution of times.

Solution

129 times were recorded.
← Any relevant information perhaps not covered in location, spread and shape?

An estimate for the mean time is 37.6 minutes.
The median time lies in the 25 to 35 minute class.
← Comment about location. Mean, median (whichever can be determined).

The times were spread out from about 5 minutes to 105 minutes, i.e. the range was about 100 minutes. An estimate for the standard deviation (σ_n) is 23.4 minutes
← Comment about spread. Range, interquartile range, standard deviation (whichever can be determined)

15 – 25 minutes is the modal class (which is just the second of the ten classes).
The long tail to the right indicates that the distribution is positively skewed.
← Comment on shape. Symmetry, modality, skewness, as appropriate.

Whilst the times ranged from approx 5 minutes to 105 minutes over half (55%) were between 5 minutes and 35 minutes.
← Anything else you notice of relevance.

Descriptions could also include mention of: gaps, clusters, more dense/less dense regions, extreme values or outliers.

EXAMPLE 3

Compare the distributions shown in the box plots below.

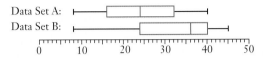

Solution

The median for data set B, 36, is much higher than that of data set A, 24.
Both data sets have a lowest score of 8 but set B has the greater highest score, 45, compared to 40 for set A.

← Compare location.

Data set B has a range of 37 compared to 32 for set A.
Both data sets have an interquartile range of 16.

← Compare spread.

The box plot for set A is symmetrical and each quarter of the scores span 8 marks. On the other hand the longer left whisker of set B and the greater part of the box being to the left of the median suggest the set B marks are skewed to the left.

← Compare shape.

The median score in set A is the same as the lower quartile score in set B.
The top 25% of the marks from set B exceeded the top mark in set A.

← Anything else of relevance.

Exercise 4B

Describe each of the distributions shown in questions 1 to 6.

1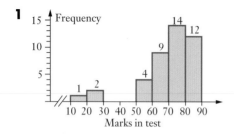

2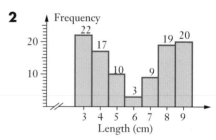

3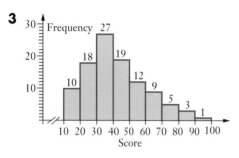

4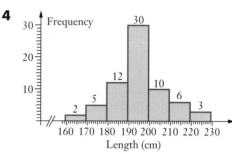

5

Score	31 − 35	36 − 40	41 − 45	46 − 50	51 − 55	56 − 60	61 − 65
Frequency	15	12	16	15	13	14	15

6

Score (x)	$0 \leq x < 10$	$10 \leq x < 20$	$20 \leq x < 30$	$30 \leq x < 40$	$40 \leq x < 50$	$50 \leq x < 60$	$60 \leq x < 70$
Frequency	46	29	13	6	3	2	1

7 Compare the distributions shown in the box plots below.

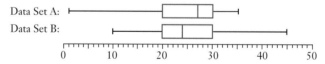

8 Compare the distributions shown in the box plots below which were formed using the results of two maths classes, set A and set B, taking the same test, with set A being the top set and expected to do better than set B which was the second set.

Each boxplot shows any outliers that are more than

$$1.5 \times \text{the interquartile range from the nearest quartile}$$

as separate crosses.

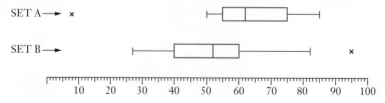

ISBN 9780170390262

9 The rainfall figures recorded at a Regional Meteorology Station for each day that some rain fell at the location, from 1 January to 31 December of a particular year, are shown in the table and graph below:

Rainfall (x mm)	$0 < x < 5$	$5 \leq x < 10$	$10 \leq x < 15$	$15 \leq x < 20$	$20 \leq x < 25$	$25 \leq x < 30$	$60 \leq x < 65$
Number of days	67	14	18	3	6	1	1

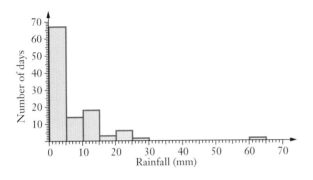

Write a report describing the rainfall for this region in the year for which the data applies.

10 The table below shows the scores obtained by the 196 students sitting a particular examination.

Score	21 to 30	31 to 40	41 to 50	51 to 60	61 to 70	71 to 80	81 to 90	91 to 100	101 to 110	111 to 120
Number of students	2	3	11	12	17	32	37	41	24	17

Write a summary describing the performance of the students in this examination.

11 A survey of the age of the donors and the recipients of a particular organ transplant procedure led to the following histograms:

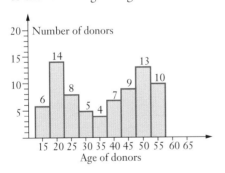

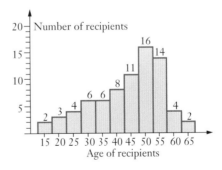

Write a report describing and comparing the data.

Miscellaneous exercise four

This miscellaneous exercise may include questions involving the work of this chapter, the work of any previous chapters, and the ideas mentioned in the Preliminary work section at the beginning of the book.

1 Find the mean and the standard deviation (correct to two decimal places) of the following set of scores both with and without the outlier included.

$$4, \quad 5, \quad 6, \quad 6, \quad 8, \quad 9, \quad 9, \quad 9, \quad 34.$$

2 The diagram on the right shows boxplots for percentage scores in three tests taken by a maths class.

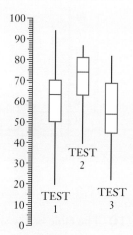

 a In which test was the highest mark scored?

 b In which test was the lowest mark scored?

 c Which test had the highest median?

 d Which test had the greatest interquartile range?

 e Which test had the greatest range of marks?

 f Which test had the smallest range of marks?

 g What percentage of the students scored 50% or above in test one?

 h In which test did at least three quarters of the students achieve a mark of more than 60%?

3 If five of the scores from the central column of the histogram shown on the right were removed from the data, would the standard deviation increase or decrease? Justify your answer with appropriate reasoning.

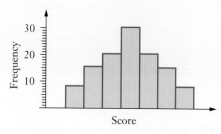

4 The pie chart below left shows (for a particular year) the intended destinations for the following year for the year 12 students in Western Australia. The pie chart below right shows where these same students actually were the following year.

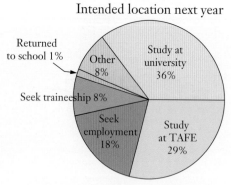

Intended location next year

Returned to school 1%
Other 8%
Study at university 36%
Seek traineeship 8%
Seek employment 18%
Study at TAFE 29%

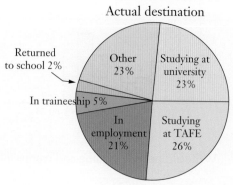

Actual destination

Returned to school 2%
Other 23%
Studying at university 23%
In traineeship 5%
In employment 21%
Studying at TAFE 26%

[Source of data: Western Australian Department of Education and Training.]

ISBN 9780170390262

Imagine you are a newspaper reporter asked to write a short article using some of the information contained in these pie charts. Concentrate on just one or two sectors, e.g. university study or TAFE study, and write the article with an appropriate headline included.
(There were approximately 18 000 year 12 students in WA that year.)

5 Given that the four sets of data that were used to create the histograms below were also used to create the four boxplots shown, match each histogram with its corresponding boxplot.

Histogram A

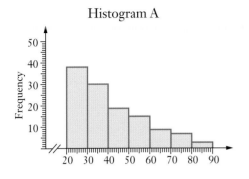

Histogram B

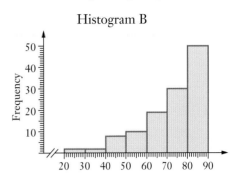

Histogram C

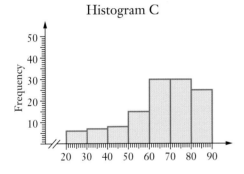

Histogram D

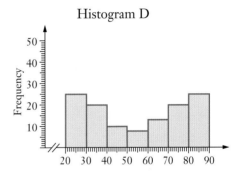

Boxplot 1

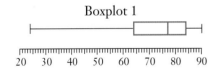

Boxplot 2

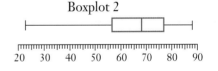

Boxplot 3

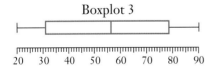

Boxplot 4

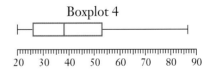

6 Classifying an outlier as:

> *any score that is more than 1.5 × the interquartile range*
> *from either quartile 1 or quartile 3, whichever is the nearer*

which scores in the following distributions would be classified as outliers?

a A distribution with a lower quartile of 28 and upper quartile of 40.

b A distribution with a median of 35, which is 5 above the lower quartile and 13 below the upper quartile.

c

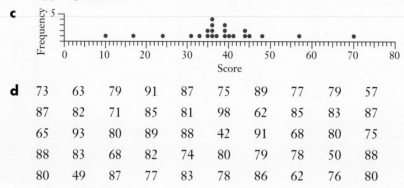

d

73	63	79	91	87	75	89	77	79	57
87	82	71	85	81	98	62	85	83	87
65	93	80	89	88	42	91	68	80	75
88	83	68	82	74	80	79	78	50	88
80	49	87	77	83	78	86	62	76	80

7 A test involved ten questions and was sat by 120 students. Copy and complete the following table showing the marks obtained. (Each question was either correct, 1 mark, or incorrect, 0 marks.)

- Whilst you may not be familiar with the term *cumulative frequency* used in the third row of the table, with thought you should be able to determine what it means.

Mark	0	1	2	3	4	5	6	7	8	9	10
Frequency	3	0	5				15	21	18		
Cumulative frequency	3	3	8	15	28	42				113	120

a How many students scored a mark of 9?

b How many students scored a mark less than 10?

c What percentage of students scored a mark greater than 7?

d What fraction of students scored 3 or less?

e Display the data as a boxplot.

f Display the data as a frequency histogram.

g Describe the distribution.

8 Find, correct to two decimal places, the standard deviation of the five numbers:

$$(a-4), \quad (a-2), \quad (a+1), \quad (a+3), \quad (a+7).$$

ISBN 9780170390262

5.

The statistical investigation process

- Implementing the statistical investigation process
- Miscellaneous exercise five

Suppose we were asked to investigate

if a particular species of animal is in decline,

if cutting a lawn makes it grow faster,

if sugary drinks cause hyperactiveness,

if social media networks are addictive,

if man-made pollution is causing global warming,

if females have a quicker reaction time than males,

if female year 8s are better at mental mathematics than male year 8s,

if year 11s are fitter than year 8s,

etc, etc

We could implement the **statistical investigation process**, a process that can be described by the following steps:

1   Clarify the problem and formulate one or more questions that can be answered with data.

2   Design and implement a plan to collect and obtain appropriate data.

3   Select and apply appropriate graphical or numerical techniques to analyse the data.

4   Interpret the results of this analysis and relate the interpretation to the original question and communicate findings in a systematic and concise manner.

In this chapter, you will be invited to carry out a statistical investigation that will involve collecting and comparing data across two or more groups to investigate a question.

For example, consider the following question:

Are year elevens fitter than year eights?

To investigate this question we would need to collect data about a group of year elevens and about a group of year eights and compare the findings.

For example, consider the following question:

Do females have quicker reaction times than males?

To investigate this question we would need to collect data about a group of males and about a group of females and compare the findings.

Both examples involve collecting data across two groups.

iStock.com/t_kimura

Implementing the statistical investigation process

Collect data to investigate one of the questions posed below (or negotiate with your teacher to collect data that will involve collecting and comparing data across two or more groups to investigate an alternative question of your own).

You need to think about   what data you will collect,
                              how you will collect it,
                              how you will record it.

Present your findings as a report and include in your report:

- the question you decided to investigate,

- what data you collected, how you collected it and how you recorded it,

- the collected data, tabulated and/or graphed as appropriate,

- your analysis of that data,

- your conclusions.

Are year 11s the fittest year group in the school?

                                 Do females have quicker reaction times to males?

Are female year 8s better at mental arithmetic than male year 8s?

                           How does your *Mathematics Applications* class compare
                      with other *Mathematics Applications* classes in recent tests?

Do males have bigger heads than females?

                          Are left-handers generally taller than right-handers?

Miscellaneous exercise five

This miscellaneous exercise may include questions involving the work of this chapter, the work of any previous chapters, and the ideas mentioned in the Preliminary work section at the beginning of the book.

1 The times taken to perform a particular task were recorded for 174 people and the results, in seconds, are shown below:

20.41	30.13	60.34	19.72	81.31	27.14	54.14	70.08	68.73	17.24
51.34	20.32	26.83	84.72	25.12	41.13	24.23	53.24	31.62	67.81
26.14	57.24	29.81	70.02	76.24	55.82	58.21	24.55	24.62	44.14
60.82	26.08	63.54	20.80	10.73	40.21	27.23	64.18	75.64	53.80
29.24	48.24	85.34	57.89	38.13	24.32	39.23	75.00	35.88	23.14
74.18	80.79	11.23	33.15	77.34	68.81	28.14	24.21	16.81	31.54
79.31	38.72	88.23	70.73	69.14	78.85	69.86	70.13	76.88	75.21
18.74	70.54	33.65	38.94	37.24	29.53	48.87	30.82	23.22	45.61
76.31	38.24	56.18	66.84	60.98	62.41	15.71	49.00	64.92	31.70
38.16	48.94	47.32	18.31	39.41	28.32	77.41	39.91	34.73	67.34
74.23	60.93	85.23	74.89	34.13	75.32	61.24	75.87	74.98	23.91
33.24	29.11	36.40	65.11	20.81	51.24	13.75	43.24	29.18	35.24
47.41	71.34	89.10	71.23	76.71	28.71	71.92	67.14	58.90	53.71
79.58	42.09	61.82	85.67	65.01	75.93	61.34	58.32	31.23	47.24
21.32	77.23	88.24	14.63	76.91	42.90	34.89	23.62	53.28	72.81
57.31	18.13	65.23	76.52	48.21	73.24	78.51	23.51	42.64	15.23
42.71	79.22	42.83	36.02	62.13	22.04	30.24	51.89	31.24	64.99
77.14	29.72	21.41	81.25						

The histogram for these results, using the intervals

$10 \le$ time < 30 $\qquad$ $30 \le$ time < 50 $\qquad$ $50 \le$ time < 70 $\qquad$ $70 \le$ time < 90

is shown below.

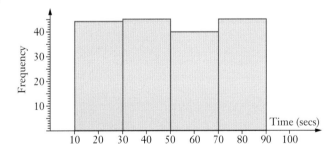

This histogram suggests that the times are evenly, or *uniformly*, distributed from a low of about 10 seconds to a high of about 90 seconds.

However, is this 'uniform distribution' really the case?

Draw the histogram for this data using the intervals

$10 \le$ time < 20 $\qquad$ $20 \le$ time < 30 $\qquad$ $30 \le$ time < 40 $\qquad$ etc.,

and comment on your findings.

2 Expand and simplify each of the following:

    a $2(x + 4) + 5(x + 3)$ **b** $5(2x + 3) + 2(3x + 1)$

    c $12(x + 1) - 5(x + 2)$ **d** $3(x - 4) - 2(x - 1)$

    e $3(2x + 5) - 1(x - 3)$ **f** $2(3x + 2) - (5x + 2)$

3 The box plot and the histogram for a set of data are shown on the right.

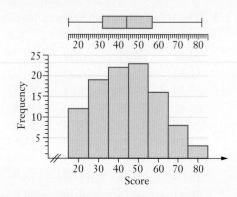

For this set of data, determine

    a   the lowest score,

    b   the median,

    c   the highest score,

    d   the interquartile range,

and use your calculator to determine

    e   an estimate for the mean,

    f   an estimate for the standard deviation.

4 Concerned about the spread of a certain disease, the health department launches a long-term campaign aimed at increasing the population's awareness of the risks. To monitor the success of the campaign amongst various sections of the community, all newly diagnosed sufferers are noted according to a number of characteristics.

The graphs below show the percentage of newly diagnosed sufferers in each category, before the campaign started and 5 years into the campaign.

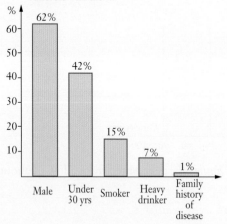

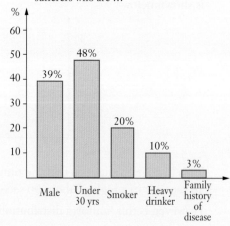

    a   The percentages for each graph add up to more than 100%! Explain.

    b   Would it be correct to say that the graphs indicate that the number of newly diagnosed sufferers who are male has decreased but the number who are under 30 has increased? Justify your answer.

6.

Solving
equations

- Solving equations
- Equations with brackets or fractions
- Linear equations
- Miscellaneous exercise six

The *Preliminary work* section at the beginning of this book reminded us of work encountered previously involving

> *using a formula to determine the value of a variable, or pronumeral, that appears in the formula by itself and on one side of the equals sign, given the values of the variables, or pronumerals appearing on the other side.*

In the examples given, there we used $A = P + I$ to determine A, given P and I,

$$C = 2\pi r \text{ to determine } C \text{ knowing } \pi \text{ and given } r,$$

$$s = ut + \frac{1}{2}at^2 \text{ to determine } s, \text{ given } u, a \text{ and } t.$$

Suppose instead, in this last formula, we were asked to determine u given s, t and a.

Substituting the given values into the formula gives us a statement of equality, called an **equation**, which will have to be **solved** to determine the unknown quantity.

For example, if $s = 63$, $t = 3$ and $a = 10$, $s = ut + \frac{1}{2}at^2$ becomes

$$63 = u(3) + \frac{1}{2}(10)(3)^2$$

i.e. $63 = 3u + 45$

Solving this **equation** means finding the value of u for which $3u + 45$ does equal 63.

What is the required value of u for which

$$3u + 45 = 63?$$

Note: The word *variable* was used in chapter one when considering data. In that case there were *various* responses that could be given to a question like 'What is your favourite colour?' Hence the use of the word variable there.

In the formula $A = P + I$, A can take *various* values, dependent upon the values of P and I. Hence the use of the word variable in this situation. Given specific values of P and I, the particular value of A can be determined.

Solving equations

To solve $3u + 45 = 63$, did you proceed mentally,
    or did you perhaps use the ability of some calculators to solve equations
    or did you carry out a step-by-step process to isolate u?

These three approaches are shown below and in the examples that follow.

Solving $3u + 45 = 63$ mentally:

We know that *eighteen* plus forty-five equals 63.
Thus $\quad 3u = 18$.
But three multiplied by *six* is eighteen.
Hence $\quad u = 6$.

Using the solve facility on a calculator.

    solve (63=3·u+45, u)
                        {u=6}

Using a step-by-step process of doing something to both sides of the equation in order to isolate the
unknown whilst retaining the correctness of the equality statement.

Starting with the given equation: $\qquad 63 = 3u + 45$
We subtract 45 from each side to isolate $3u$: $\quad 18 = 3u$
We divide each side by 3 to isolate u: $\qquad 6 = u$
Thus $\qquad\qquad\qquad\qquad\qquad\qquad\quad u = 6.$

EXAMPLE 1

Solve the following equations

a $\quad x + 9 = 21$ $\qquad\qquad$ **b** $\quad 5x - 7 = 23$ $\qquad\qquad$ **c** $\quad 15 - 2x = 4$

Solution

a $\quad x + 9 = 21$

Mentally:
We know that *twelve* add nine equals twenty-one.
Thus $x = 12$.

Using the solve facility:

    solve (x+9=21, x)
                    {x=12}

YAY Media AS / Alamy Stock Photo

Step by step approach to isolate x:

We are given the equation: $\qquad x + 9 = 21$

We subtract 9 from each side to isolate x: $\qquad x = 21 - 9$

$\qquad\qquad\qquad\qquad = 12$

Thus $\qquad x = 12$

b $\quad 5x - 7 = 23$

Mentally:

We know that *thirty* take seven equals twenty-three.

Thus $5x = 30$.

But five times *six* equals 30 and so $x = 6$.

Using the solve facility:

solve (5x−7=23, x)

$\qquad\qquad\qquad\qquad$ {x=6}

Step by step approach to isolate x:

We are given the equation: $\qquad 5x - 7 = 23$

We add 7 to both sides to isolate $5x$: $\qquad 5x = 23 + 7$

$\qquad\qquad\qquad\qquad = 30$

Now we divide both sides by 5 to isolate x: $\qquad x = 30 \div 5$

$\qquad\qquad\qquad\qquad = 6$

Thus $\qquad x = 6$

c $\quad 15 - 2x = 4$

Mentally:

We know that fifteen take *eleven* equals four.

Thus $2x = 11$ and so $x = 5.5$.

Using the solve facility:

solve (15−2x=4, x)

$\qquad\qquad\qquad\qquad$ {x=5.5}

Step by step approach to isolate x:

We are given the equation: $\qquad 15 - 2x = 4$

We add $2x$ to both sides to make the x term positive: $\qquad 15 = 4 + 2x$

We subtract 4 from both sides to isolate $2x$: $\qquad 15 - 4 = 2x$

$\qquad\qquad \therefore \qquad 11 = 2x$

Now we divide both sides by 2 to isolate x: $\qquad 5.5 = x$

$\qquad\qquad \therefore \qquad x = 5.5$

More complex
equations

Equations with brackets or fractions

Some equations may involve brackets. For example, $3(2x + 1) - 5 = 40$.

Some equations may involve fractions. For example, $\dfrac{2x + 3}{5} = 4$.

EXAMPLE 2

Solve the following equations

a $3(x - 1) = 21$

b $3(2x + 1) - 5 = 40$

c $\dfrac{2x + 3}{5} = 4$

Solution

a $3(x - 1) = 21$

> **Mentally:**
> We know that three times *seven* equals twenty-one.
> Thus $x - 1 = 7$.
> But *eight* take one equals 7 and so $x = 8$.

> **Using the solve facility:**

> solve (3(x−1)=21, x)
>
> {x=8}

> **Step by step approach to isolate *x*:**
> We are given the equation: $3(x - 1) = 21$
> Expand to remove bracket: $3x - 3 = 21$
> Add 3 to both sides to isolate $3x$: $3x = 24$
> Now divide both sides by 3 to isolate x: $x = 8$

b $3(2x + 1) - 5 = 40$

> **Mentally:**
> We know that three *fifteens* take away five is equal to 40.
> Thus $2x + 1 = 15$.
> But *fourteen* add one is equal to fifteen.
> Hence $2x = 14$ and so $x = 7$.

ISBN 9780170390262

Using the solve facility:

solve (3(2x+1)−5=40, x)
{x=7}

Step by step approach to isolate x:

We are given the equation:	$3(2x + 1) - 5 = 40$
Expand to remove bracket:	$6x + 3 - 5 = 40$
Hence	$6x - 2 = 40$
Add 2 to both sides to isolate $6x$:	$6x = 42$
Now divide both sides by 6 to isolate x:	$x = 7$

c $\dfrac{2x + 3}{5} = 4$

Mentally:

We know that *twenty* divided by five is equal to four.

Thus $2x + 3$ must equal 20.

But *seventeen* plus three is equal to twenty.

Hence $2x = 17$ and so $x = 8.5$.

Using the solve facility:

solve $\left(\dfrac{2 \cdot x + 3}{5} = 4,\ x\right)$
{x=8.5}

Step by step approach to isolate x:

We are given the equation:	$\dfrac{2x + 3}{5} = 4$
Multiply both sides by 5 to remove fractions:	$2x + 3 = 20$
Subtract 3 from both sides to isolate $2x$:	$2x = 17$
Now we divide both sides by 2 to isolate x:	$x = 8.5$

iStock.com/PeopleImages

If the equations are more involved the mental approach can be too difficult but we can still use either

- the solve facility on a graphic calculator, or
- the step by step process to isolate the unknown,

as shown below.

EXAMPLE 3

Solve the following equations:

a $2(x + 3) - 3(2x + 1) = -5$

b $\dfrac{x}{7} = \dfrac{3}{10} + \dfrac{2x}{21}$

Solution

a **Using the solve facility:**

> solve (2(x+3)−3(2x+1)=−5, x)
> $\qquad\qquad\qquad\qquad$ {x=2}

Step by step approach to isolate x:

We are given the equation:	$2(x + 3) - 3(2x + 1) = -5$
Expand to remove brackets:	$2x + 6 - 6x - 3 = -5$
Collect like terms:	$-4x + 3 = -5$
We add $4x$ to both sides to make the x term positive:	$3 = -5 + 4x$
Add 5 to both sides to isolate $4x$:	$8 = 4x$
Now we divide both sides by 4 to isolate x:	$2 = x$
	$\therefore \quad x = 2$

b **Using the solve facility:**

> solve $\left(\dfrac{x}{7} = \dfrac{3}{10} + \dfrac{2x}{21}, x \right)$
> $\qquad\qquad\qquad\qquad$ {x=6.3}

Step by step approach to isolate x:

We are given the equation:

$$\dfrac{x}{7} = \dfrac{3}{10} + \dfrac{2x}{21}$$

Multiply both sides by 210 to remove fractions: $\quad 210 \times \dfrac{x}{7} = 210 \times \dfrac{3}{10} + 210 \times \dfrac{2x}{21}$

i.e.: $\qquad\qquad\qquad\qquad\qquad\qquad\qquad 30x = 63 + 20x$

Subtract $20x$ from each side: $\qquad\qquad\qquad 10x = 63$

Now we divide both sides by 10 to isolate x: $\qquad x = 6.3$

EXAMPLE 4

A firm manufacturing a particular motorbike determines that the profit, P, made from the production and sale of x of these bikes is given by

$$P = 5400x - 238\,000$$

Calculate the number of these bikes the firm must produce and sell to make a profit that exceeds one million dollars.

Solution

Either use the ability of some calculators to determine the value of x when $P = 1\,000\,000$ in the given formula.

```
Equation:
P=5400·x−238000

○ P=1000000
⦿ x=229.259259259259
Lower=−9E+999
Upper=9E+999
```

Or substitute $P = 1\,000\,000$ into the given formula and use the step by step approach to determine x, as shown below.

We are given the formula: $\qquad\qquad P = 5400x - 238\,000$

Substitute $P = 1\,000\,000$: $\qquad 1\,000\,000 = 5400x - 238\,000$

Add $238\,000$ to both sides to isolate $5400x$: $\quad 1\,238\,000 = 5400x$

$$\therefore \qquad x \approx 229.3$$

However the situation requires x to take positive integer values and so:

The firm must produce and sell at least 230 of these bikes to make a profit that exceeds one million dollars.

EXAMPLE 5

Formula: $s = \dfrac{(u+v)}{2}t$.

Find t given that $s = 35$, $u = 10$ and $v = 4$.

Solution

By calculator:

```
Equation:
s=u+v/2·t

○ s=35
○ u=10
○ v=4
⦿ t=5
Lower=−9E+999
Upper=9E+999
```

Substitute and then isolate t:

$$s = \frac{(u+v)}{2}t$$

$$35 = \frac{(10+4)}{2}t$$

$$\therefore \quad 35 = 7t$$

$$\therefore \quad t = 5$$

Formula: $S = 2\pi r^2 + 2\pi rh$

Find h given that $S = 545$ and $r = 5$ giving your answer correct to 1 decimal place.

Solution

Either
Determine the value of h for the given values of S and r using the ability of some calculators to determine unknown values in a formula given sufficient information.

Equation:
S=2·π·r²+2·π·r·h

○ S=545
○ r=5
◉ h=12.3478887970166
Lower=−9E+999
Upper=9E+999

Or
Substitute $S = 545$ and $r = 5$ into the given formula and use the step by step approach to determine h, as shown below.

We are given the formula: $S = 2\pi r^2 + 2\pi rh$

Substitute $S = 545$ and $r = 5$: $545 = 2\pi(5)^2 + 2\pi(5)h$

$\therefore\quad 545 = 50\pi + 10\pi h$

Subtract 50π from both sides to isolate $10\pi h$: $545 - 50\pi = 10\pi h$

Now we divide both sides by 10π to isolate h: $\dfrac{545 - 50\pi}{10\pi} = h$

$\therefore\quad h = 12.3$ (to one decimal place)

Thus when $S = 545$ and $r = 5$, $h = 12.3$ (to one decimal place).

Linear equations

Equations code puzzle

The six equations shown below are all examples of **linear equations in one variable**.

$$2x + 17 = 5 \qquad\qquad 5p - 7 = 32 \qquad\qquad 15 - 3z = 6$$

$$\frac{w}{2} = 5 \qquad\qquad 2(3q - 5) + 1 = 15 \qquad\qquad \frac{2n + 3}{5} = 4$$

Each equation, after expansion of any brackets and separation of fractions, only involves terms that are either just a number, or the variable multiplied or divided by a number. Linear equations do *not* involve the variable squared (x^2), cubed (x^3), square rooted ($\sqrt{x}$), in the denominator of a fraction $\left(\dfrac{3}{x-1}\right)$, as a power ($2^x$) etc.

Each of the above equations can, with a bit of work, be written in the form $ax + b = 0$, the basic form of a linear equation in one variable:

$$2x + 12 = 0 \qquad\qquad 5p - 39 = 0 \qquad\qquad -3z + 9 = 0$$
$$w - 10 = 0 \qquad\qquad 6q - 24 = 0 \qquad\qquad 2n - 17 = 0$$

(As we will see later in this text, and as you may already be familiar with, equations of the form $y = ax + b$ give straight line, or **linear**, graphs.)

The emphasis in this text will be on solving **linear** equations.

Exercise 6A

(Use this exercise to practise the various methods shown in the previous pages.)

1 Solve the following equations.

a $x + 5 = 11$

b $5 - x = 31$

c $x + 3 = 31$

d $3x + 7 = 25$

e $15 - 2x = 6$

f $3x - 7 = 2$

g $2(x + 3) = 14$

h $3(x - 1) = 21$

i $5(x + 2) = 15$

j $2(x - 5) = 16$

k $3(1 + x) = 18$

l $5(2x - 1) = 9$

m $\dfrac{x}{3} = 5$

n $\dfrac{x}{2} = 21$

o $\dfrac{3x}{10} = 1.5$

p $\dfrac{5x}{7} = 1$

q $\dfrac{x}{7} = 12$

r $\dfrac{3x - 5}{2} = 8$

s $3(x + 2) + 5(2x - 1) = 27$

t $7(2x + 3) - 3(2x + 1) = 10$

u $\dfrac{x}{2} + 5 = 11$

v $\dfrac{x}{7} = \dfrac{6}{21}$

w $\dfrac{4x}{3} + \dfrac{3}{4} = -\dfrac{x}{6}$

x $\dfrac{x}{2} - \dfrac{2x - 1}{5} = 2$

y $\dfrac{3x + 1}{5} + \dfrac{5x - 1}{4} = 24$

2 $A = P + I$

  a Find P given that $A = 676$ and $I = 26$.

  b Find A given that $P = 1250$ and $I = 85$.

  c Find I given that $P = 1185$ and $A = 1240$.

3 $v = u + at$

  a Find v given that $u = 14$, $a = 2$ and $t = 3$.

  b Find u given that $v = 30$, $a = 3$ and $t = 4$.

  c Find a given that $v = 30$, $u = 16$ and $t = 7$.

  d Find a given that $v = 8$, $u = 20$ and $t = 4$.

  e Find t given that $v = 23$, $u = 5$ and $a = 6$.

  f Find t given that $v = 12$, $u = -10$ and $a = 2$.

4 $C = 2\pi r$

  a Find r given that $C = 25$. (Answer correct to two decimal places.)

  b Find r given that $C = 95$. (Answer correct to two decimal places.)

  c Find C given that r $= 8$. (Answer correct to two decimal places.)

  d Find r given that $C = 128\pi$.

5 $A = 2\pi rh$

  a Find A given that $r = 1$ and $h = 4$. (Answer correct to 2 decimal places)

  b Find r given that $A = 125$ and $h = 7$. (Answer correct to 2 decimal places)

  c Find h given that $A = 200$ and $r = 6$. (Answer correct to 2 decimal places)

6 $s = \dfrac{(u + v)}{2} t$

  a Find t given that $s = 56$, $u = 3$ and $v = 5$.

  b Find u given that $s = 92$, $v = 14$ and $t = 8$.

  c Find v given that $s = 22.5$, $u = -6$ and $t = 5$.

7 If three resistors, R_1, R_2 and R_3 are placed in an electrical circuit as shown in the diagram they are said to be in **series**. The total resistance, R, is then given by:

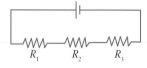

$$R = R_1 + R_2 + R_3$$

  a Find R if $R_1 = 2$, $R_2 = 5$, $R_3 = 4$.

  b Find R if $R_1 = 10$, $R_2 = 15$, $R_3 = 20$.

  c Find R_1 if $R = 32$, $R_2 = 8$, $R_3 = 14$.

  d Find R_3 if $R = 80$, $R_1 = 25$, $R_2 = 28$.

8 $S = 2\pi r^2 + 2\pi rh$

  a Find h given $S = 250$ and $r = 4$. (Answer correct to two decimal places.)

  b Find h given $S = 1000$ and $r = 4$. (Answer correct to two decimal places.)

9 A certain mass of gas is held under pressure at a constant temperature. V, the volume of the gas, is related to the pressure according to the rule:

$$V \times P = 150.$$

  a Find V given that $P = 6$,

  b Find P given that $V = 15$,

  c Find V given that $P = 7.5$,

  d Find P given that $V = 60$.

10 When a forklift truck lifts an object of mass m kg through a height h metres it gives the object potential energy equal to P Joules where $P = 9.8\,mh$.

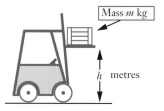

  a Find P when $m = 50$ and $h = 1.5$.

  b Find P when $m = 400$ and $h = 2$.

  c Find m when $P = 2940$ and $h = 2$.

  d Find h when $P = 13\,230$ and $m = 600$.

11 A car rental firm charges $\$C$ for renting a particular vehicle where C depends on d, the number of days hired, and k, the number of kilometres travelled. C is calculated according to the rule:
$C = 30 + 70d + 0.1k$.

  a Find the cost of renting the vehicle for 7 days and travelling 500 km.

  b Find the cost of renting the vehicle for 3 days and travelling 1200 km.

  c A person wishes to rent the vehicle for six days but does not want the hire costs to exceed $\$700$. How far could the person travel in the car in the six days?

12 When an agent of a particular real estate company sells a house for A thousand dollars the commission the agent receives is P where

$$P = 600 + 4A$$

a Find the amount the agent receives in commission for the sale of each of the following houses:

$625\,000 \qquad\qquad \$700\,000 \qquad\qquad \$375\,000

b An agent sells a house and receives a commission of $2380. How much was the house sold for?

c An agent sells a house and receives a commission of $4000. How much was the house sold for?

13 If, in an archaeological dig, human bones are found, or if in a macabre murder case parts of a body are discovered, these bones can be used to estimate the height the person was when alive. In particular the bone in the upper arm from elbow to shoulder, called the humerus, is a good indicator of height.

If the humerus is of length h cm then a reasonable estimate for the height of a male with this length humerus is:

$$(2.9h + 71) \text{ cm,}$$

and a reasonable estimate for the height of a female with this length humerus is:

$$(2.75h + 71) \text{ cm.}$$

a In an archaeological dig the remains of a male are uncovered and the humerus is found to be 34 cm long. Estimate the height of the male.

b What would be the expected humerus length of a 1.81 metre tall female?

14 A company has five thousand calendars printed. If it sells x of these calendars, where x is from a low of zero to a high of 5000, the profit produced will be P where P is given by:

$$P = 12.7x - 29\,750$$

a What will be the profit if the firm sells 2500 of the calendars?

b What will be the profit if the firm sells 3500 of the calendars?

c What will be the profit if the firm sells all but 800 of the calendars?

d What is the least number of the calendars the firm needs to sell to make a profit that exceeds $9000?

e If the firm are left with 3200 of the calendars unsold determine whether they have made a profit or a loss and state how much?

f What is the greatest profit the firm can make from this venture?

g If the firm sold none of the calendars how much would they lose?

h What is the least number of calendars the firm must sell to avoid making a loss?

Miscellaneous exercise six

This miscellaneous exercise may include questions involving the work of this chapter, the work of any previous chapters, and the ideas mentioned in the Preliminary work section at the beginning of the book.

1 Solve the following equations.

    a $3x - 1 = 20$ **b** $5x + 6 = 24 - 4x$ **c** $3x + 7 = 21 - x$

    d $6(2x + 1) - 5 = 31$ **e** $\dfrac{x - 2}{3} - 3 = 1$ **f** $\dfrac{2x}{9} = \dfrac{5}{3}$

2 Formula: $A = 4bh + b^2$

    a   Find A given that $b = 3$, $h = 5$.

    b   Find h given that $A = 119$, $b = 7$.

3 The following nine scores are listed in ascending order, from left to right.

    $a + 2,$ $a + 3,$ $b,$ $c - 1,$ $c - 1,$ $c + 1,$ $d,$ $e - 1,$ $a + e.$

    The box and whisker diagram for these nine scores is as follows:

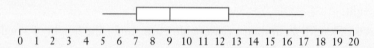

    Determine a, b, c, d and e.

4 To pass a particular course a student needs to gain a mean of at least 60% in the six tests that form the course assessment. In the first five tests the student achieves marks of 65%, 58%, 71%, 60% and 59%. What percentage mark does the student require in test six in order to pass the course?

5 The box plots on the right are for four sets of data, A, B, C and D. Which of the data sets:

    a   seems to involve the greatest variability?

    b   has the smallest interquartile range?

    c   has the smallest range?

    d   contains the lowest of all the scores?

    e   could the following apply to?

       More than half of the scores in set __ exceed all of the scores in set __.

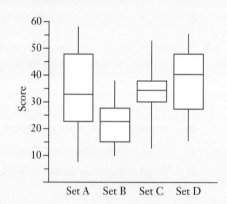

6 The marks obtained by 47 students in an examination marked out of 140 are shown below:

$$
\begin{array}{cccccccccccc}
25 & 32 & 50 & 54 & 59 & 66 & 67 & 69 & 71 & 73 & 75 & 75 \\
76 & 76 & 78 & 80 & 81 & 82 & 83 & 83 & 84 & 85 & 88 & 88 \\
89 & 89 & 89 & 89 & 90 & 92 & 94 & 95 & 96 & 99 & 100 & 104 \\
106 & 108 & 109 & 109 & 111 & 111 & 113 & 114 & 115 & 115 & 117 &
\end{array}
$$

With the mean of these marks being $\bar{x}$ and the standard deviation σ, grades are awarded to these 47 students as follows

		exam mark	$\geq$	$\bar{x} + 1.25\sigma$	Grade A
$\bar{x} + 0.5\sigma$	$\leq$	exam mark	$<$	$\bar{x} + 1.25\sigma$	Grade B
$\bar{x} - 0.5\sigma$	$\leq$	exam mark	$<$	$\bar{x} + 0.5\sigma$	Grade C
$\bar{x} - 2\sigma$	$\leq$	exam mark	$<$	$\bar{x} - 0.5\sigma$	Grade D
		exam mark	$<$	$\bar{x} - 2\sigma$	Fail grade

Determine the number of students obtaining each grade.

7 A real estate agent in a particular region wants to publish the average price of houses sold in the town each month. Past figures indicate that each month somewhere between 10 and 50 houses are sold each month. Most houses in the area are of a similar nature except for a small number of beachside luxury properties. These do not come up for sale very often but when they do they are priced very much above most others in the region.

Which average (mean, median or mode) would you advise the agent to use for the average monthly price and why?

8 The salaries of the 187 full time employees of a large manufacturing company were distributed as follows:

Category	Salary $S	Number of Employees
A	$40\,000 \leq S < 50\,000$	23
B	$50\,000 \leq S < 60\,000$	64
C	$60\,000 \leq S < 70\,000$	43
D	$70\,000 \leq S < 80\,000$	25
E	$80\,000 \leq S < 90\,000$	14
F	$90\,000 \leq S < 100\,000$	9
G	$100\,000 \leq S < 110\,000$	5
H	$110\,000 \leq S < 120\,000$	3
I	$120\,000 \leq S < 130\,000$	1

a Calculate the mean and standard deviation of this distribution.

b With increased automation in the manufacturing processes the company no longer requires such a large workforce. Through voluntary redundancy and non-replacement of retirees the company reduced its workforce to 173 by losing 5 employees from category A, 4 from category B, 3 from category C and 2 from category F. Calculate the mean and standard deviation of the salaries of this workforce of 173.

9 One hundred and twenty-seven people applied for a particular job vacancy and the company involved decided to invite what they considered to be the best 50 applicants to take an aptitude test. On test day 47 of the invited 50 turned up and the marks obtained (out of 75) were as shown in the following diagram:

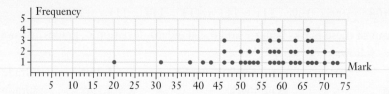

The company decides to invite for interview any of the applicants who achieved a mark in the test that is more than 1 standard deviation above the mean. How many of the applicants do they invite for interview?

10 Describe each of the following distributions.

a

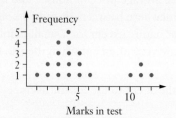

b

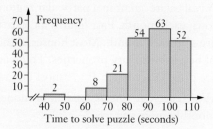

11 The graphs below show the age distribution of the human population of two countries, A and B.

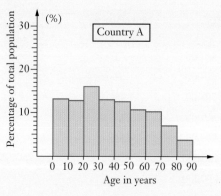

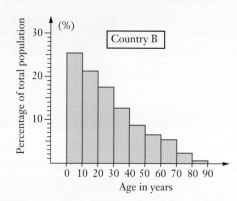

a Which of the two countries has the greater population?

b If country A has a total population of approximately eighty two million people, approximately how many of these are 70 or over?

c Produce a report describing the population distributions of each country, commenting on any similarities and differences between the distributions of the two countries and comment on possible implications for future government policy given these population distributions.

ISBN 9780170390262

7.

Using equations to solve problems

- Pyramids
- Number puzzles
- Solving problems
- Equations from simple interest formula
- Equations from ratios
- Miscellaneous exercise seven

Pyramids

The pyramid pattern shown on the right is to be completed by adding the 2 and the 5 and putting the answer in the box indicated by the arrows, adding the 5 and the 7 and putting the answer in the box indicated by the arrows etc.

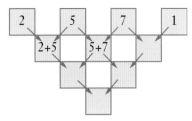

The completed pyramid is shown on the right.

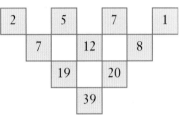

The pyramids shown below all follow this style. Copy and complete each pyramid.

1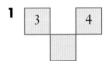

2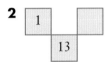

3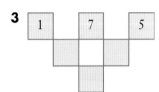

4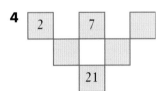

5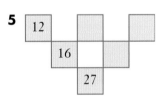

6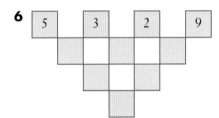

7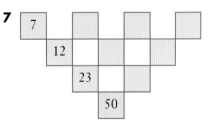

Pyramids 1 to 7 should not have caused you too much trouble but now try pyramids 8 to 14 below. They are not so easy and may involve fractions.

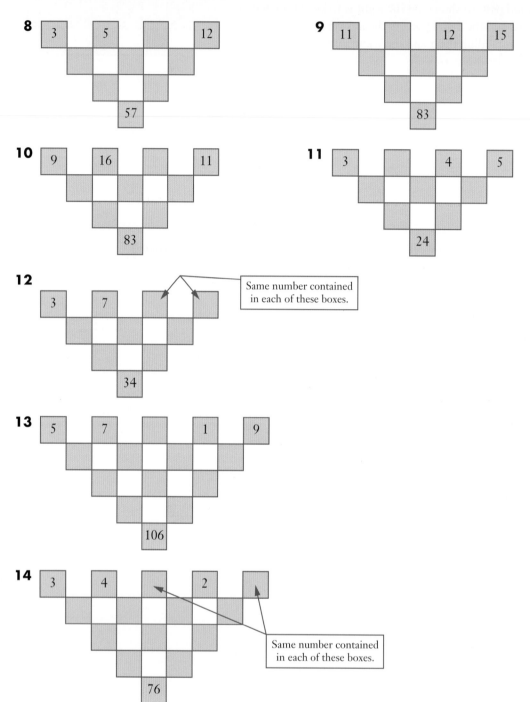

8
| 3 | | 5 | | | | 12 |

57

9
| 11 | | | | 12 | | 15 |

83

10
| 9 | | 16 | | | | 11 |

83

11
| 3 | | | | 4 | | 5 |

24

12
| 3 | | 7 | | | | |

Same number contained in each of these boxes.

34

13
| 5 | | 7 | | | | 1 | | 9 |

106

14
| 3 | | 4 | | | | 2 | | |

Same number contained in each of these boxes.

76

How did you get on with the pyramids on the previous pages? Did you develop any techniques for finding the missing numbers in pyramids 8 to 14?

When asked to complete a pyramid like the one below, a common initial reaction is 'we don't have enough information'. In fact, as you probably found out from the similar questions on the previous page, there is enough information but the empty box in the first line can make it seem that we cannot 'get started'.

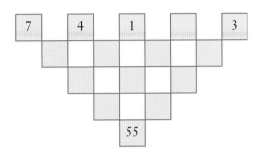

To overcome this 'getting started' difficulty, we could use trial and adjustment. For example, we could try some number, say 2, in the empty box in the first row and complete the pyramid using this number. If the last box is then less than 55, our initial guess needs to be increased. If the last box is greater than 55, then our initial guess needs to be decreased. Thus we 'get started' by trying a number and then, based on information this number produces, we adjust and improve our initial trial. Hence the name **trial and adjustment**.

Alternatively we could overcome the 'I can't get started' problem by using a symbol, for example a letter, to represent the unknown number in the top row of the pyramid.

Following through the pyramid using this symbol, say x, we obtain the entry for the last box in terms of x.

Thus, for this pyramid, the last box, $32 + 4x$, must equal 55.

Thus $32 + 4x = 55$.

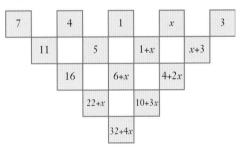

Solving this equation by one of the methods of the previous chapter, i.e. mentally, using the solve facility on a calculator or using a step by step process to isolate x gives the solution to this equation as

$$x = 5.75$$

The pyramid can then be completed:

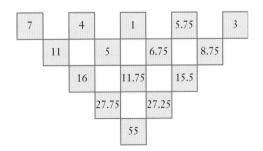

This technique of introducing a symbol, frequently x, is a useful mathematical technique for solving questions in which we seem to have sufficient information but we can't seem to 'get started'.

Exercise 7A

Find the value of x in each of the following pyramids.

1
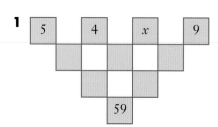

2
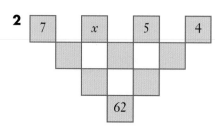

3
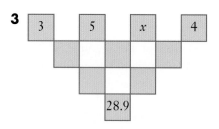

4
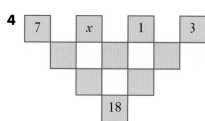

5
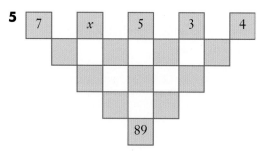

6
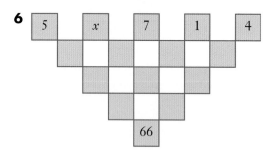

7 The 'hexapatterns' shown below **all** follow the pattern shown on the right. The symbols ☆, ◆ and ✳ all represent numbers.

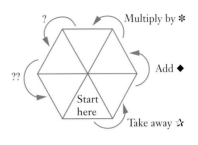

Multiply by ✳

Add ◆

Take away ☆

?

??

Start here

a What number does ☆ represent?

b What number does ◆ represent?

c What number does ✳ represent?

d What instruction should go in place of '?' ?

e What instruction should go in place of '??' ?

f Copy and complete the 'hexapatterns' E to L shown below.

A 15, 16, 5, 8, 0, 1

B 18, 20, 6, 10, 1, 2

C 24, 28, 8, 14, 3, 4

D 312, 412, 104, 206, 99, 100

E 5

F 2.5

G 16

H 132

I 25

J 44

K 40

L 18

Number puzzles

So far in this chapter we have seen an x introduced to enable us to 'get started' on certain problems. Once started we then obtained an equation which allowed x to be determined. This technique of introducing an x in order to get started can be used to solve other types of puzzles and problems.

Word problems with equations

iStock.com/pamspix

EXAMPLE 1

I think of a number, multiply it by three, add seven to the answer and then add the number I first thought of. If my answer is 51 what number did I think of?

Solution

Question   We need to multiply the number by three and then add seven etc. but how can we get started if we do not know what the number is?

Answer   To allow us to get started let x be the number first thought of.

Let x be the number first thought of:	x
Multiply by 3:	$3x$
Add seven to the answer:	$3x + 7$
Add the number first thought of:	$3x + 7 + x$
The result is 51:	$\therefore \quad 3x + 7 + x = 51$
Collect like terms:	$4x + 7 = 51$

Solving: Step-by-step   OR   By calculator

$$4x + 7 = 51$$
$$4x = 51 - 7$$
$$4x = 44$$
$$x = 11$$

solve(4x+7=51, x)
$\{x=11\}$

The number I first thought of was 11.

Note: In the last example we did not leave the answer as $x = 11$. The initial problem asked us to find the number thought of and had no mention of x which we introduced to help solve the question. Our final sentence should clearly answer this question and not simply say '$x = 11$' unless of course the question itself had introduced, and asked us to find, x.

EXAMPLE 2

I think of a number, subtract it from twelve, multiply the result by five, add the number I first thought of and end up with 24. Find the number first thought of.

Solution

Let x be the number first thought of:	x
Subtract it from 12:	$12 - x$
Multiply by 5:	$5(12 - x)$
Add the number first thought of:	$5(12 - x) + x$
The result is 24:	$\therefore \quad 5(12 - x) + x = 24$

Solving:

$$5(12 - x) + x = 24$$
$$60 - 5x + x = 24$$
$$60 - 4x = 24$$
$$60 = 24 + 4x$$
$$36 = 4x$$
$$\therefore \quad x = 9$$

solve(5(12–x)+x=24, x)
$\{x=9\}$

The number I first thought of was 9.

Note: Trial and adjustment is a perfectly acceptable alternative method for solving this type of question. For example 2:

Try 3: Subtract 3 from 12:   9
Multiply by 5:   45
Add number first thought of: 48
Thus 3 was not correct as we want to end up with 24.

Try 5: Subtract 5 from 12:   7
Multiply by 5:   35
Add number first thought of: 40
Thus 5 was not correct but was better that 3.

Continuing in this way will eventually give us the correct answer of 9. Trial and adjustment is a very useful technique but it can be a lengthy process, particularly if the answer is not an integer. In the examples that follow the calculator approach or the step by step method will tend to be shown.

EXAMPLE 3

I think of a number, add five, multiply the result by four, and end up with an answer that is 29 more than the number I first thought. Find the number first thought of.

Solution

Let x be the number first thought of.

Thus $\qquad 4(x + 5) = x + 29$ ◄— Check carefully that you understand how this equation has been arrived at.

Solving:

```
solve(4(x+5)=x+29, x)
                        {x=3}
```

The number I first thought of was 3.

EXAMPLE 4

I think of a number, add fifteen, divide the result by two, and end up with four more than the number I first thought. Find the number first thought of.

Solution

Let x be the number first thought of.

Thus $\qquad \dfrac{x + 15}{2} = x + 4$ ◄— Check carefully that you understand how this equation has been arrived at.

Solving:

Multiply by two: $x + 15 = 2x + 8$
Subtract x: $15 = x + 8$
Subtract 8: $7 = x$
$\qquad\qquad\qquad\therefore x = 7$

The number I first thought of was 7.

Exercise 7B

1 If x represents 'the number' write each of the following statements in terms of x.

Example: Three times the number then add one.
Answer: $3x + 1$

  a Multiply the number by five then add six.

  b Take the number away from fourteen.

  c Add six to the number and then multiply by five.

  d Double the number then take away seven.

  e Take seven from the number and then double your answer.

  f Double the number, add five and then multiply your answer by three.

2 If x is used to represent 'the number I think of' in each of the statements A to H below, select the equation from the box on the right that matches each statement and then solve the equation.

  A: I think of a number, double it and add one and my answer is 10.

  B: If I take one from the number I think of and double the result my answer is ten.

  C: I think of a number and subtract it from ten and my answer is one.

  D: I think of a number, add one and then double the result and my answer is ten.

  E: I think of a number, divide it by two and then subtract one and end up with 10.

  F: I think of a number, take ten away and my answer is one.

  G: I think of a number, subtract one and then divide by two and my answer is ten.

  H: Twice the number I thought of exceeds ten by one.

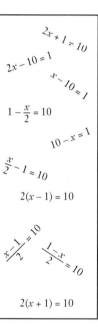

$2x + 1 = 10$

$2x - 10 = 1$

$x - 10 = 1$

$1 - \dfrac{x}{2} = 10$

$10 - x = 1$

$\dfrac{x}{2} - 1 = 10$

$2(x - 1) = 10$

$\dfrac{x - 1}{2} = 10$

$\dfrac{1 - x}{2} = 10$

$2(x + 1) = 10$

3 I think of a number, multiply by three, subtract eleven from the answer and then add the number I first thought of. If my answer is 25 what was the number I first thought of?

4 I think of a number, add seven, multiply the result by two and end up with an answer that is 17 more than the number I first thought of. Find the number I first thought of.

5 I think of a number, double it, add seven and end up with an answer that is 17 more than the number I first thought of. Find the number I first thought of.

6 I think of a number, subtract four, multiply the result by three and end up with an answer that is two less than the number I first thought of. Find the number I first thought of.

7 I think of a number, subtract five, divide the answer by two and end up with seven less than the number I first thought of. Find the number I first thought of.

8 I think of a number, add two, multiply the answer by three and then subtract eleven. This gives an answer equal to twice the number first thought of. Find the number I first thought of.

ISBN 9780170390262

9 I think of a number, double it, add three, multiply the result by two, take away the number I first thought of and end up with twenty seven. Find the number I first thought of.

10 I think of a number, multiply it by three and then subtract one. If one quarter of this answer is added to the number I first thought of the answer is 22.5. Find the number I first thought of.

11 I think of a number, subtract it from fourteen, multiply the result by three and find that my answer exceeds the number I first thought of by four. Find the number I first thought of.

12 I think of a number, double it and add three. I find that if I take twenty one from my answer I end up with half of the number I first thought of. Find the number I first thought of.

Solving problems

All of the questions of the previous exercise were of the 'think of a number' style. They could all be solved by introducing an x to represent the number. This allowed us to 'get started', form an equation and determine x. The following examples show this same technique used to solve a variety of problems.

EXAMPLE 5

Jackie has saved $28 more than John. Between them they have saved $154. How much has each person saved?

Solution

Let the amount John has saved be	$x
The amount Jackie has saved will then be	$(x + 28)
Between them they have saved $154. Thus	$x + (x + 28) = 154$
i.e.	$2x + 28 = 154$
This equation can be solved to give	$x = 63$

John has saved $63 and Jackie has saved $91.

Note: It is not the intention here to claim that the introduction of x is the only way to solve the problem. As has been mentioned before, 'trial and adjustment' can be used. Alternatively the answer can be 'reasoned through'. For example, the previous question could be solved as follows:

If we take Jackie's extra $28 from the $154 then the answer, $126, is their total if they both had John's amount. Thus John's amount must be $126 ÷ 2 = $63. Hence Jackie must have saved $63 + $28 = $91.

The arithmetic involved, i.e. taking 28 from 154 and then dividing by two, is the same as we would do to solve the equation $2x + 28 = 154$ by the step by step process, but the answer was 'reasoned through' rather than obtaining and solving an equation in x. If you choose this 'reasoning' method be sure to explain what you are doing so that others can follow your reasoning.

EXAMPLE 6

Rosalyn is 8 years older than Jennifer. In six years' time their ages will be such that Rosalyn will be twice as old as Jennifer. How old is Jennifer now?

Solution

Let Jennifer's age now be	x years
Thus Rosalyn's age now is	$(x + 8)$ years
In 6 years' time Jennifer will be	$(x + 6)$ years
In 6 years' time Rosalyn will be	$(x + 8) + 6$ years
Thus	$(x + 8) + 6 = 2(x + 6)$
This equation can be solved to give	$x = 2$

Jennifer is 2 years old now.

Note: Alternatively 'trial and adjustment' can be used or the answer can be 'reasoned through' as follows: If Rosalyn is 8 years older than Jennifer now, she will always be 8 years older. Thus, in 6 years' time, when Rosalyn is twice as old as Jennifer, the difference in their ages will be both 8 years and one lot of Jennifer's age. Thus in 6 years' time Jennifer will be 8 and Rosalyn will be 16. Thus Jennifer is 2 years old now.

EXAMPLE 7

An amateur drama group hire a theatre for their production. They expect to sell all of the 1200 tickets, some at $10 and the rest at $7. The group require the ticket sales to cover their $4150 production costs, to allow a donation of $4000 to be made to charity and to provide a profit of $1000 to aid future productions. If they are to exactly achieve this target and their expectations regarding ticket sales are correct how many of the total 1200 tickets should they charge $10 for and how many should they charge $7 for?

iStock.com/&#169 Pavel Losevsky

Solution

Let the number of $7 tickets be	x
These will give an income of	$7x$ dollars
The number of $10 tickets will then be	$(1200 - x)$
These will give an income of	$10(1200 - x)$ dollars
Thus	$7x + 10(1200 - x) = 4150 + 4000 + 1000$
which can be simplified to	$12\,000 - 3x = 9150$
Solving gives	$x = 950$

The group should sell 950 tickets at $7 each and 250 tickets at $10 each.

Exercise 7C

1 Tony and Bob each put some money towards the purchase of a new car they need for their business. Tony puts in $5500 more than Bob. Together they put in a total of $18 500. How much does each contribute?

2 Three people, Sue, Lyn and Paul run a part time business. At the end of their first year they decide that the profits should be shared out such that Lyn gets one and a half times as much as Sue, and Paul gets $5000 more than Sue. If the profits for the year are $47 000 how much should each receive?

3 Bill is 29 years older than his daughter Rebecca. In fifteen years' time their ages will be such that Bill will be twice as old as Rebecca. How old is Bill now?

4 A manufacturer sells a particular product for $40 per unit. The manufacturer's costs for producing the units consist of a fixed $5000 plus a cost of $22 per unit. If the manufacturer produces and sells x units find,

  a an expression in terms of x for the cost to the manufacturer for producing these x items,

  b the value of x for the manufacturer to at least 'break even'. (Assume that x must take integer values.)

5 A firm manufactures two types of chair, the deluxe and the standard. In one week the firm manufactures a total of 120 of these chairs, x of the standard and $(120 - x)$ of the deluxe. Each standard chair requires 3 hours of work and each deluxe requires 4 hours of work.

  a Find an expression in terms of x for the number of hours required to make the x standard chairs.

  b Find an expression in terms of x for the number of hours required to make the $(120 - x)$ deluxe chairs.

  c If the 120 chairs required 405 hours of work altogether find how many of each type of chair were made.

6 A farmer wishes to fence off a rectangular area with the length 10 metres longer than the width. If the farmer has 360 metres of fencing available for this task what should be

  a the width of the rectangle?

  b the length of the rectangle?

  c the area of the rectangle?

iStock.com/bluecinema

7 If Heidi's current age in years is multiplied by five and two is subtracted from the answer the result is equal to her Dad's age in years. If Heidi is currently x years old find an expression for her Dad's age.

In 8 years' time the ages will be such that Heidi's Dad will be three times as old as Heidi. How old is Heidi now?

8 An amateur drama group hire a theatre for their production. They expect to sell all 850 tickets, some at $12 and the rest at $8. The group require the ticket sales to cover their $3760 production costs and to make a profit of $4000. If they are to exactly achieve this target and their expectations regarding ticket sales are correct how many of the 850 tickets should they charge $12 for and how many should they charge $8 for?

9 A farmer has a certain number of acres that she wishes to use to grow wheat, barley and lupins.

Whatever acreage she decides to use for the lupins she likes to have 2000 acres more than this for barley. She also likes to use twice as many acres for wheat as she does for barley.

a If she uses x acres for lupins, find

 i an expression in terms of x for the number of acres she uses for barley,

 ii an expression in terms of x for the number of acres she uses for wheat.

b If she decides to use a total of 18 000 acres for the three products, determine how many acres she uses for each.

10 A firm making fertiliser produces a new fertiliser *QuickGrow*. Each 50 kg bag of *QuickGrow* contains x kg of compound X and $(50 - x)$ kg of compound Y. Each kilogram of X contains 150 g of a particular element and each kilogram of Y contains 80 g of this element. The company wants each 50 kg of *QuickGrow* to contain 6.24 kg of this element.

a Find

 i an expression in terms of x for the amount of the particular element contained in x kg of compound X and state the units,

 ii an expression in terms of x for the amount of the particular element contained in $(50 - x)$ kg of compound Y and state the units.

b How much of each compound, X and Y, should each 50 kg of *QuickGrow* contain to give the desired total amount of the particular element?

11 A book shop owner orders some hardback and some softback versions of a book. The hardback version costs the shop owner $20 each and the softback $12 each. The total order was for 300 books.

When the order arrives there were only 200 books! The shopkeeper wishes to query the order but cannot find his copy of the original order. However his records do tell him that it was going to cost him $5080. How many of each type of the book was his original order for?

iStock.com/David Hanlon

12 An investor has $5000 to invest and decides to invest some of it with company A and the rest with company B. After one year each $1 invested with company A has grown to $1.20, each dollar invested with company B has grown to $1.05 and the $5000 has grown to $5670.

How much of the original $5000 was invested with each company?

ISBN 9780170390262

Equations from simple interest formula

In Unit One of this *Mathematics Applications* course you would have determined the *simple interest*, I, earned when P is invested for T years in an account paying $R\%$ per annum simple interest.

The formula used in this situation is: $$I = \frac{PRT}{100}$$

If instead we use R in decimal form we use $I = PRT$. For example if the interest rate is 6% the first formula would use $R = 6$ but the second would use $R = 0.06$.

In Unit One we were determining I given P, R and T. If instead we want to determine P, or R or T, given I and the other two, we can substitute values into the appropriate formula and solve the resulting equation.

EXAMPLE 8

How much money needs to be invested for 3 years at 6% simple interest to earn interest of $864?

Solution

Using $\qquad I = PRT$
Given $I = 864$, $R = 0.06$, $T = 3$ then
$\qquad 864 = P \times 0.06 \times 3$
i.e. $\qquad 864 = 0.18P$
and so $\qquad P = \dfrac{864}{0.18}$
$\qquad\qquad = 4800$
The investment needs to be $4800.

EXAMPLE 9

How long does an investment of $12 500 need to be invested at 3.8% per annum simple interest to earn interest of $2375?

Solution

Using $\qquad I = PRT$
Given $P = 12\,500$, $R = 0.038$, $I = 2375$
then $\qquad 2375 = 12\,500 \times 0.038 \times T$
i.e. $\qquad 2375 = 475T$
and so $\qquad T = \dfrac{2375}{475}$
$\qquad\qquad = 5$
The investment needs to be for 5 years.

Note

Alternatively, questions like the previous two examples could be solved using the built-in capability of some calculators to perform simple interest calculations.

Equations from ratios

The *Preliminary work* at the beginning of this text reminded us of the idea of a ratio. In particular the following example was given:

Suppose the ratio of males to females in a school is 17 : 21.

If we know that there are 231 females in the school we can determine the number of males

males : females = 17 : 21 $\searrow$ × 11
$\qquad\qquad\qquad$ =   ? : 231

The number of males = 17 × 11, i.e. 187.

$$\boxed{\begin{array}{ll} 231 \,/\, 21 & \\ & 11 \end{array}}$$

Alternatively, if we let the number of males be m, we could set up, and solve, an equation, as follows:

$$\text{males : females} = 17 : 21$$

$\therefore\qquad\qquad\qquad\qquad\qquad m : 231 = 17 : 21$

Hence $\qquad\qquad\qquad\qquad\qquad \dfrac{m}{231} = \dfrac{17}{21}$

× by 231 to eliminate fractions: $\qquad 231 \times \dfrac{m}{231} = 231 \times \dfrac{17}{21}$

$$m = 231 \times \dfrac{17}{21}$$

$$= 187$$

The number of males is 187, as before.

Suppose the previous situation had instead given us the ratio as:

$$\text{females : males} = 21 : 17.$$

We would then have proceeded as follows:

$\therefore\qquad\qquad\qquad\qquad\qquad 231 : m = 21 : 17$

Hence $\qquad\qquad\qquad\qquad\qquad \dfrac{231}{m} = \dfrac{21}{17}$

Now with the variable in the denominator this is not a linear equation but this need not trouble us. We could solve the equation by:

- first multiplying by $17m$ to eliminate the fractions, as shown below left, or

- using the fact that if $\dfrac{a}{b} = \dfrac{c}{d}$, then it follows that $\dfrac{b}{a} = \dfrac{d}{c}$, as shown below right.

Given	$231 : m = 21 : 17$	OR	Given	$231 : m = 21 : 17$
	$\dfrac{231}{m} = \dfrac{21}{17}$			$\dfrac{231}{m} = \dfrac{21}{17}$
× by $17m$:	$17m \times \dfrac{231}{m} = \dfrac{21}{17} \times 17m$		Hence	$\dfrac{m}{231} = \dfrac{17}{21}$
Hence	$17 \times 231 = 21m$		× by 231:	$231 \times \dfrac{m}{231} = \dfrac{17}{21} \times 231$
Solving gives	$m = 187$		Hence	$m = 187$

MATHEMATICS APPLICATIONS Unit 2 ISBN 9780170390262

Notice that if
$$\frac{a}{b} = \frac{c}{d}$$

Then multiplying by bd: $bd \times \dfrac{a}{b} = bd \times \dfrac{c}{d}$

gives $ad = bc$

Notice that this final statement could have been obtained by 'cross multiplying' the original 'fraction equals fraction' equation.

$$\frac{a}{b} \bowtie \frac{c}{d}$$

Cross multiplying is a useful shortcut but it needs to be used with care and with an understanding of why it works. Only use it in situations that are 'single fraction = single fraction'.

EXAMPLE 10

Find the values of a and b in the following

a $a : 4 = 7 : 20$

b $15 : 4b = 1 : 2$

Solution

a   Given $a : 4 = 7 : 20$

then $\dfrac{a}{4} = \dfrac{7}{20}$

Hence $20 \times a = 7 \times 4$

$a = \dfrac{28}{20}$

$= 1.4$

b   Given $15 : 4b = 1 : 2$

then $\dfrac{15}{4b} = \dfrac{1}{2}$

Hence $15 \times 2 = 4b \times 1$

$b = \dfrac{30}{4}$

$= 7.5$

In your study of unit one of this course you would have encountered the idea of **similar triangles**. Situations involving determining unknown lengths in similar triangles can often involve setting up and then solving a statement involving ratios, as the next two examples show.

EXAMPLE 11

In the diagram shown on the right AB = 5 m, BC = 9 m, and BE = 3 m. Find the length of CD, justifying your answer.

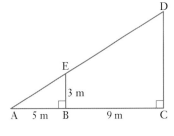

Solution

In triangles ABE and ACD: ∠EAB = ∠DAC (same angle)

∠EBA = ∠DCA (= 90°)

Hence the third angles will be equal and so △ABE ~ △ACD, corresponding angles equal.

Hence   AB : AC = BE : CD

Letting the length of CD be x m: $5 : 14 = 3 : x$

i.e. $\dfrac{5}{14} = \dfrac{3}{x}$

Hence $5x = 42$

and so $x = 8.4$

CD is of length 8.4 m.

EXAMPLE 12

So how tall is the street light?

On a sunny day, at the same time that a street light standing on horizontal ground casts a shadow of length 9.1 metres, a 1.6 metre stick held vertically on the same ground casts a shadow of length 2.6 metres.

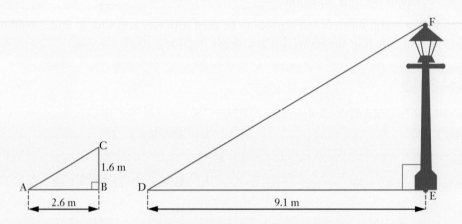

What does this information suggest that the height of the street light is?

Solution

In triangles ABC and DEF: ∠CAB = ∠FDE (angle Sun's rays make with the ground)

∠CBA = ∠FED (= 90°)

Hence the third angles will be equal and so △ABC ~ △DEF, corresponding angles equal.

Hence   AB : DE = BC : EF

Let the height of the street light be h metres.

$$2.6 : 9.1 = 1.6 : h$$

i.e. $\dfrac{2.6}{9.1} = \dfrac{1.6}{h}$

Hence $2.6h = 9.1 \times 1.6$

and so $h = 5.6$

The information suggests that the height of the street light is 5.6 metres.

Exercise 7D

1 Find the values of $a, b, c, \ldots i$ in the following.

a $a : 9 = 2 : 3$ **b** $b : 10 = 2 : 5$ **c** $4 : 3 = 8 : c$

d $d : 3 = 5 : 2$ **e** $2e : 9 = 4 : 5$ **f** $f : 2 = 7 : 5$

g $6 : g = 4 : 1$ **h** $17 : 2h = 5 : 1$ **i** $5 : 8 = 3 : i$

2 Determine how much money needs to be invested for 3 years at 8% per annum simple interest to accrue interest of $1008.

3 What annual rate of simple interest would cause an investment of $6400 to grow to $7360 in two years?

4 Sally invests a sum of money into an account paying 8.2% simple interest. After three years the account has earned interest of $209.10. What was the initial sum invested?

5 For how long must an investment of $10 000 be invested in an account paying 8.6% per annum simple interest for it to become $13 870?

6 How many days must a sum of $8650 be left to accrue interest at a rate of 5% per annum simple interest to become $8823?

7 What annual rate of simple interest is needed to see an initial investment of $6720 become $7011.20 in 8 months?

8 $65 000 is invested in an account paying simple interest at a rate of R% per annum. Nine months later the account is closed and the total balance of principal plus interest is then $68 997.50. Find R.

9 What annual rate of simple interest is required to see an investment of eight and a half thousand dollars grow to eight thousand eight hundred and six dollars in 219 days?

10 The ratio of males to females in the audience for a particular event was 5 : 7. If there were 1045 males in the audience, how many females were in the audience?

11 Suppose the ratio of male students to female students in a school is 15 : 17. If there are 345 male students in the school, how many female students are there in the school?

12 **So how tall is the tree?**

At the same time as a tree, stood on horizontal ground, casts a shadow of length 24 metres, a 1.8 metre stick held vertically on the same ground casts a shadow of length 3.2 metres.

What does this information suggest that the height of the tree is?

(Your working should include justification that triangles assumed similar are indeed similar.)

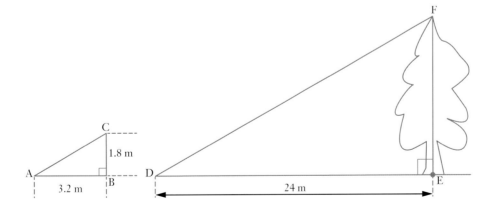

13 So how tall is the building?

When a vertical signpost that is known to have a height of 3 metres casts a 1.8 metre shadow a nearby building casts a shadow of length 15 metres.

How tall is the building?

(Your working should include justification that triangles assumed similar are indeed similar.)

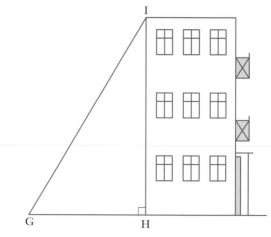

14 So how tall is the flag pole? (I)

A stick of length 2.1 metres held vertically on horizontal ground casts a shadow of length 0.6 metres.

At the same time, a nearby flagpole casts a shadow of length x metres.

If the height of the flagpole is h metres find an expression for h in terms of x.

If $x = 1.5$ find h.

(Your working should include justification that triangles assumed similar are indeed similar.)

iStock.com/George Clerk

15 So how tall is the flag pole? (II)

Standing on level ground, and with the sun shining, Panji starts at the base of a flagpole and walks along the flagpole's shadow until the tip of his own shadow is at the same point on the ground as the tip of the shadow cast by the flagpole.

Panji, who's own height is 1.75 metres, is then 2 metres from the tip of his shadow and 6 metres from the base of the flagpole.

How tall is the flagpole?

(Your working should include justification that triangles assumed similar are indeed similar.)

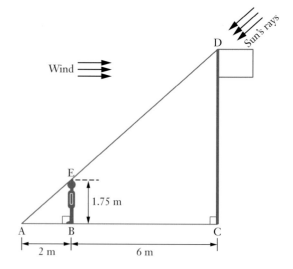

16 So how wide is the river?

As part of an initiative test a team of trainee soldiers is set the task of estimating the width of a river, from one side of the river. To do this they note a tree at the water's edge on the far side (point B in the diagram on the right) and put a pole at point A on 'their' river bank, directly opposite B. They then locate some suitable point C on the bank on their side and a point D as shown. Standing at D and looking at the tree on the opposite bank allows them to locate, and mark, point E.

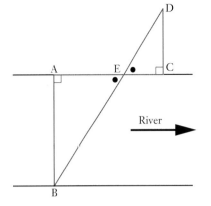

They then measure AE as being 34.6 metres, EC as 19.2 metres and CD as 29.8 metres.

How wide is the river?

(Your working should include justification that triangles assumed similar are indeed similar.)

17 So how tall is the pylon?

Roz places a small flat mirror on level ground between herself and an electricity pylon. Roz notices that with herself, the mirror and the pylon in line she can see the top of the pylon in the mirror when she is 3.5 metres from the mirror and the mirror is 17.5 metres from the middle of the base of the pylon. If Roz's eye height is 172 centimetres, how tall is the pylon?

(Your working should include justification that triangles assumed similar are indeed similar.)

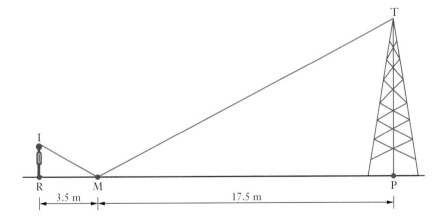

18 Mary takes out a loan which involves simple interest charged at the rate of 8% per annum. After 18 months Mary repays $1344 which clears the loan and interest.

How much did Mary borrow in the first place?

19 Mai takes out a loan which involves simple interest charged at the rate of 5% per annum. After 219 days Mai repays $7004 which clears the loan and interest.

How much did Mai borrow in the first place?

THINK OF A NUMBER

Think of a whole number between one and ten (and remember it).

Double it

Add 3

Double your answer

Add on the number you first thought of

Add four

Divide by 5

Add 1

Take away the number you first thought of.

Your answer is 3 (or at least it should be if you have followed the instructions correctly!)

Why does the above puzzle always work no matter what whole number between one and ten is chosen as the starting number? Write an explanation of why the puzzle works.

Does it work for numbers other than whole numbers between one and ten? Explain.

A TRICK INVOLVING FIFTEEN COINS

Ranii placed fifteen coins on the table and said to her sister Jenna,

'While I shut my eyes you put some of the coins in your left hand, the rest in your right hand, and then put your hands behind your back so that I cannot see them.'

Jenna followed the instructions and then Ranii said,

'Now multiply the number of coins you have in your left hand by two, add on the number of coins you have in your right hand and tell me the answer.'

Again Jenna followed the instructions and announced,

'That makes 22.'

Performing some simple arithmetic Ranii was quickly able to announce,

'You have 7 in your left hand and 8 in your right.'

'That's right,' exclaimed Jenna, *'How did you know that?'*

How did she know it? **Explain what she did and why it works.**

Miscellaneous exercise seven

This miscellaneous exercise may include questions involving the work of this chapter, the work of any previous chapters, and the ideas mentioned in the Preliminary work section at the beginning of the book.

1 Suppose the ratio of male students to female students in a school is 21 : 19. If there are 720 students in the school altogether how many female students are there in the school?

2 Find the value of x in each of the following.

a $\dfrac{x}{3} = \dfrac{7}{5}$ **b** $\dfrac{x}{10} = \dfrac{3}{4}$ **c** $\dfrac{5}{x} = \dfrac{4}{7}$ **d** $\dfrac{2}{x} = \dfrac{5}{7}$

3 Find m if $5(m + 3) + 2(3 - 2m) = 36$.

4 The mean of ten numbers is 10.8. Seven of these numbers have a mean of 12 and the other three numbers are a, $(a + d)$ and $(a + 2d)$. Determine $(a + d)$.

5 A number of adults, some male and some female, were asked ten questions about a particular issue. The number answered correctly by the adults are shown in the frequency distribution table below and, below that is the frequency table for the females in the group.

	Frequency distribution for entire group										
Number correct	0	1	2	3	4	5	6	7	8	9	10
Frequency	3	1	4	12	20	25	25	32	16	14	13

	Frequency table for the females in the group										
Number correct	0	1	2	3	4	5	6	7	8	9	10
Frequency	0	0	0	10	17	10	8	9	7	12	13

a How many females were asked the ten questions?
b How many males were asked the ten questions?
c Determine the range of the female scores.
d Determine the range of the male scores.
e Naomi claims that:

> *The range of the male scores is bigger than the range of the female scores and therefore the male scores are more spread out than the female scores.*

Comment on her claim.

6 I think of a number, multiply it by two, add one and then double the result. To this answer I add on half of the number I first thought of and end up with 83. Find the number I first thought of.

7 To the nearest $50, how much needs to be invested into an account paying 6% simple interest for the account to be worth at least $19 000 in 5 years?

8 A company invests $80 000, some into an account paying 6.3% per annum simple interest and the remainder in an account paying 5.4% per annum simple interest. After 2 years the $80 000 had grown to $89 612.

How much went into each account?

9 A maths exam was sat by 2145 candidates. The exam was marked out of 120 and the marks gained were distributed as follows.

Class interval	1 – 20	21 – 40	41 – 60	61 – 80	81 – 100	101 – 120
No. of candidates	113	263	340	720	478	231

a What is the midpoint of the 61 – 80 class?

b Use the midpoint of each interval to determine a mean and standard deviation for this grouped distribution.

10 The scores achieved by two classes in a maths test are given below:

Class One

39	33	35	44	5	37	40	24	41	30	42
12	46	52	16	58	35	22	44	37	26	28
40	50	31

Class Two

41	45	48	40	24	47	42	37	44	43	39
45	49	41	51	50	43	48	45	32	36	46

Draw a box and whisker diagram for each set of results and write a brief report comparing the distributions.

11 The scores obtained by the fifty students who sat a mathematics test are shown on this boxplot:

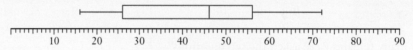

Construct a possible histogram for this set of fifty scores using the class intervals:

$$10 \le score < 20, \qquad 20 \le score < 30, \qquad 30 \le score < 40, \text{ etc.}$$

12 Two loads, load A and load B, of a particular valuable metal are for sale.

The weight of load A is greater than that of load B but load B is of a higher quality and therefore has a greater price per kilogram. The ratio of the weights of the loads is as follows:

weight of load A : weight of load B = 15 : 8.

The ratio of the per kilogram price of the two loads is as follows:

price of each kg of load A : price of each kg of load B = 3 : 5.

If the price of load A is $72 000 and the price of load B is x, find x.

ISBN 9780170390262

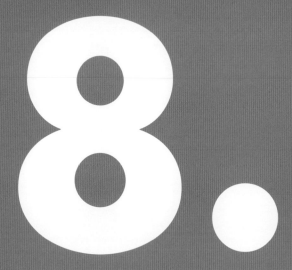

Linear relationships

- Straight line graphs
- The gradient of a straight line
- Table of values
- What's my rule?
- Table – rule – graph
- Lines parallel to the axes
- Use of a calculator with a graphing facility
- More about $y = mx + c$, the equation of a straight line
- It may not look like $y = mx + c$ but it may still be linear
- Determining the equation of a straight line
- A useful rule
- Calculator routines
- Linear relationships in practical situations
- Miscellaneous exercise eight

Situation One

Suppose that each copy of a particular book weighs 1.5 kg.

If we place copies of this book on a set of scales, each time we add one more book so the weight shown will increase by 1.5 kg, as shown in the graph on the right.

Copy and complete the following table for the situation:

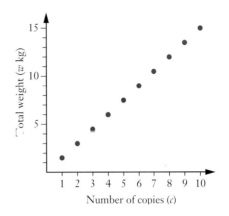

Number of copies, c	1	2	3	4	5	6	7	8	9	10
Number of kg, w										

Which of the following rules agree with the figures in your table?

$$w = 2c \qquad c = w \qquad w = 1.5c \qquad c = 1.5w \qquad c + w = 1.5$$

Situation Two

At 8 a.m. one morning there are 20 large concrete blocks in a builder's yard that need to be delivered to a worksite. A truck from a transport company is due to arrive at the builder's yard in one hour to pick up four blocks, take them to the worksite and then return for four more one hour later, repeating the process until all twenty have been removed from the builder's yard. Check that you agree that the graph on the right is consistent with this information.

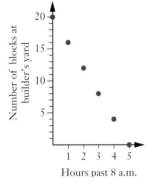

Copy and complete the following table for this situation:

Number of hours past 8 a.m., h	0	1	2	3	4	5
Number of blocks at builder's yard, n						

Which of the following rules agree with the figures in your table?

$$n = 20h \qquad n = -4h + 20 \qquad n = 4h - 20 \qquad n = 24h \qquad n = 20 - h$$

Situation Three

The graph on the right shows the amount charged by a plumber working at your house for up to 2 hours.

Copy and complete the following table for this situation:

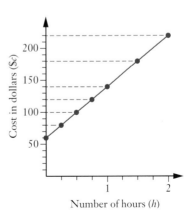

Number of hours, h	0	0.25	0.5	0.75	1	1.5	2
Cost in dollars, c							

Which of the following rules agree with your table?

$$c = 80 + 60h \qquad c = 80h + 60 \qquad c = 60 - 80h$$

Why are the dots joined in this graph and not for situations one and two?

Situation Four

Three electricians, Sparky, Flash and Voltman, have different ways of calculating a customer's bill.

- Sparky charges a standard rate per hour and has no other charges.

- Flash has a fixed, or 'standing' charge and then charges a certain amount per hour on top of that.

- Voltman has a higher standing charge than Flash but then charges less per hour.

These three methods are shown graphed below:

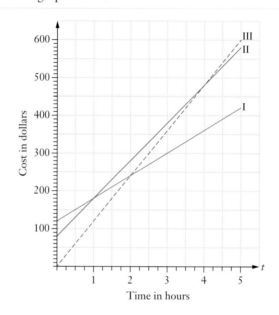

- Which line, I, II or III, corresponds to **a** Sparky, **b** Flash, **c** Voltman?

- Ignoring the standing charges who charges most per hour? What feature of the graph shows this?

- With the charge, or cost, being $C and the time being t hours the equation of line I is

$$C = 60t + 120.$$

   Determine the equations of lines II and III.

- If you were considering using one of the three electricians for a job and wanted to keep the cost to a minimum which of the three could you dismiss from your considerations?

ISBN 9780170390262

Straight line graphs

The situations on the previous two pages each gave rise to graphs for which the plotted points lay in a straight line. This is because in each situation, for each unit increase horizontally the vertical change remains constant. For example, for the first three situations:

A page of number planes

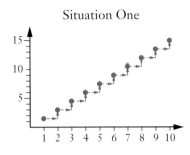

Situation One

Each time we move right 1 unit we move up 1.5 units.

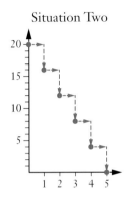

Situation Two

Each time we move right 1 unit we move down 4 units.

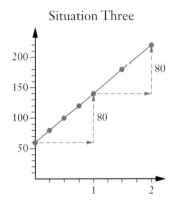

Situation Three

Each time we move right 1 unit we move up 80 units.
(Moving right one quarter of a unit sees a vertical rise of just 20 units.)

The gradient of a straight line

This vertical rise for each horizontal unit increase is called the **gradient** or **slope** of the straight line. For example:

Drawing gradients

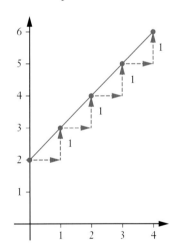

Each time we move right 1 unit we move up 1 unit.
Gradient = 1

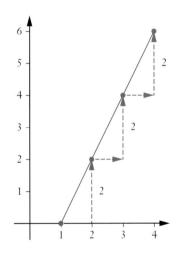

Each time we move right 1 unit we move up 2 units.
Gradient = 2

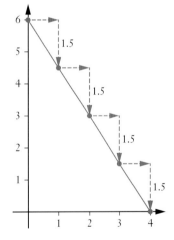

Each time we move right 1 unit we move **down** 1.5 units.
Gradient = –1.5

Note carefully the use of the negative gradient in the third case to indicate that the line moves **down** as we move to the right.

Table of values

Using points with whole number horizontal coordinates to create a table of values for the three graphs just encountered, with c as the horizontal coordinate and w the vertical coordinate, the table for the first graph (shown again on the right) is:

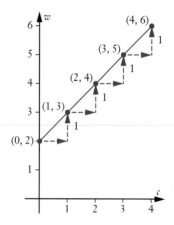

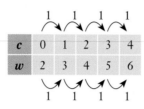

Notice how the gradient of the line, i.e. the constant increase in the w-values for each unit increase in the c-values, is evident from the table.

For the other two graphs the tables (and graphs) are shown below:

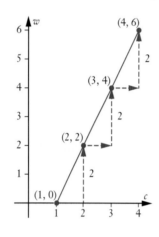 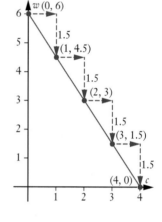

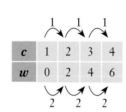 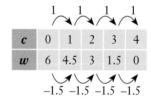

Notice again how the gradient of 2 and of –1.5 are evident from these tables.

- The gradient of a straight line graph is the vertical rise in the graph for each unit moved to the right. (A fall in the graph for each unit moved right indicates that the gradient is negative.) This gradient, or *slope*, is sometimes described as '*the rise divided by the run*'.

- With the horizontal coordinate increasing by 1 unit each time, the table of values for a straight line graph will show a constant difference pattern in the vertical coordinate equal to the gradient of the line. (If, for a constant increase in the values of the horizontal coordinate, the values of the vertical coordinate do not show a constant difference pattern the table of values is not for a straight line graph.)

Exercise 8A

For each of questions 1 to 20 determine the gradient of the given straight line.

1
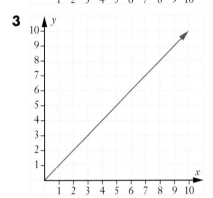

2
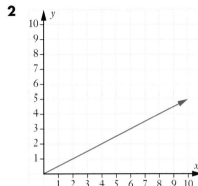

3
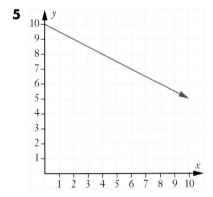

4
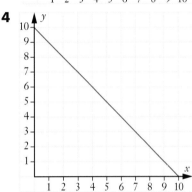

5
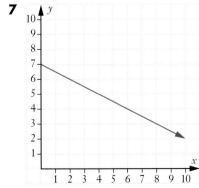

6
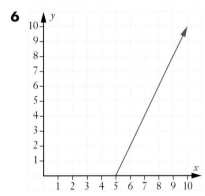

7

8
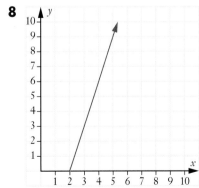

ISBN 9780170390262

9

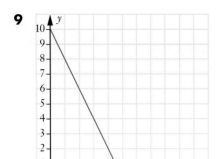

10

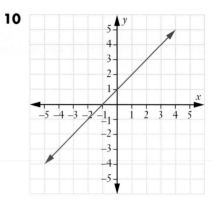

11

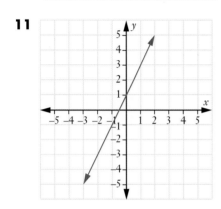

12

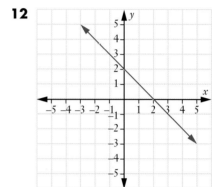

13

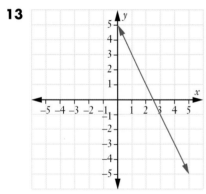

14

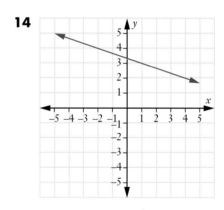

15

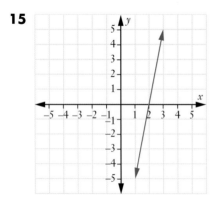

16

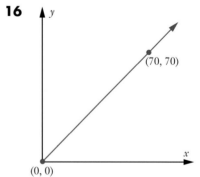

17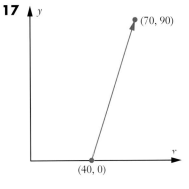

18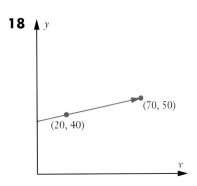

19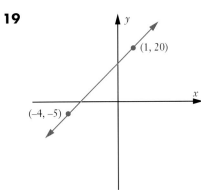

20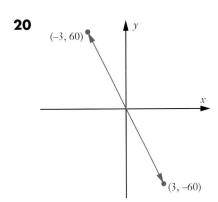

For questions 21 to 34 state whether or not the given values (x, y) would lie in a straight line if plotted and, for those cases for which a straight line would be formed, what would be the gradient of that straight line?

21

x	0	1	2	3	4	5
y	4	6	8	10	12	14

22

x	0	1	2	3	4	5
y	19	17	15	13	11	9

23

x	0	1	2	3	4	5
y	0	1	4	9	16	25

24

x	6	7	8	9	10	11
y	5	7	1	3	16	10

25

x	7	8	9	10	11	12
y	12	17	22	27	32	37

26

x	0	2	4	6	8	10
y	20	19	17	14	10	5

27

x	1	3	5	7	9	11
y	50	40	30	20	10	0

28

x	3	4	6	9	13	16
y	3	5	7	9	11	13

29

x	6	5	4	3	2	1
y	12	15	18	21	24	27

30

x	2	3	4	5	6	7
y	−17	−14	−11	−8	−5	−2

31

x	1	3	4	5	6	7
y	8	12	14	16	18	20

32

x	4	2	6	1	3	5
y	17	11	23	8	14	20

33

x	11	13	9	12	10	8
y	24	48	8	35	15	3

34

x	8	14	4	10	6	12
y	11	26	1	16	6	21

8. Linear relationships ●●●●●●●●○○○●●

What's my rule?

Gradient and y-intercept

Finding the equation of a line

Notice that the straight line graph on the right passes through the following points (as well as others):

$$(-3, -5), \quad (-2, -3), \quad (-1, -1), \quad (0, 1),$$
$$(1, 3), \quad (2, 5), \quad (3, 7).$$

For every one of these points the x and y coordinates fit the rule

$$y = 2x + 1.$$

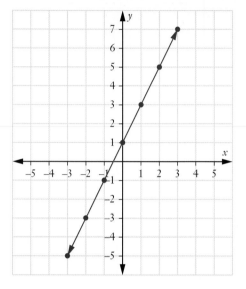

Indeed every point lying on the given straight line will have coordinates that fit the rule

$$y = 2x + 1$$

and all points not on the line will not fit the rule.

For example: The point $(1.5, 4)$ lies on the line and
$$4 = 2 \times 1.5 + 1$$
The point $(-2.5, -4)$ lies on the line and
$$-4 = 2 \times (-2 \cdot 5) + 1$$
The point $(2, 3)$ does not lie on the line and
$$3 \neq 2 \times 2 + 1$$
The point $(-3, 0)$ does not lie on the line and
$$0 \neq 2 \times (-3) + 1$$

We say that the rule for the straight line shown is $y = 2x + 1$.

Note that the line with a rule of $y = 2x + 1$ has
- a gradient of 2

and • cuts the vertical axis at $(0, 1)$.

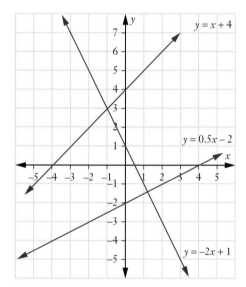

Similarly: $y = 1x + 4$ has • a gradient of 1
and • cuts the y-axis at $(0, 4)$.

$y = 0.5x - 2$ has • a gradient of 0.5
and • cuts the y-axis at $(0, -2)$.

$y = -2x + 1$ has • a gradient of -2
and • cuts the y-axis at $(0, 1)$.

To generalise:

> The straight line with the rule $\quad y = mx + c$
> has $\quad$ a gradient of m
> and $\quad$ cuts the y-axis at the point $(0, c)$.

And:

> If a straight line has gradient m and cuts the y-axis at the point $(0, c)$, it has the rule:
> $$y = mx + c$$

ISBN 9780170390262

Note: In saying that the rule for a straight line has the form $y = mx + c$ the choice of the letters m and c is not important. We could equally well have said that straight lines have rules of the form
$$y = px + k, \qquad y = ax + b, \qquad y = bx + a, \qquad y = cx + d, \qquad y = rx + s, \quad \text{etc.}$$
or as
$$y = k + px, \qquad y = b + ax, \qquad y = a + bx, \qquad y = d + cx, \qquad y = s + rx, \quad \text{etc.}$$
It is the form of the rule that is important, not the use of m and c.

However $y = mx + c$ is one of the expressions more frequently used, as is $y = a + bx$, so in this text we will tend to use one or other of these two.

For the straight line shown on the right:
Gradient = 2   (Each move of one unit to the right sees the line go up 2 units.)
The line cuts the vertical axis at $(0, -2)$.
Thus the rule for the line is $y = 2x - 2$.

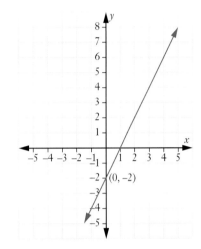

The points $(2, 2)$ and $(4, 6)$ lie on the line. Confirm that these values for x and y do indeed fit the rule.

Choose another point on the line and similarly confirm that its coordinates fit the rule.

Choose a point that does not lie on the line and confirm that its coordinates do not fit the rule.

For the second straight line shown right:
Gradient = –1.5   (Each move of one unit to the right sees the line go down 1.5 units.)
The line cuts the vertical axis at $(0, 3)$.
Thus the rule for the line is $y = -1.5x + 3$.

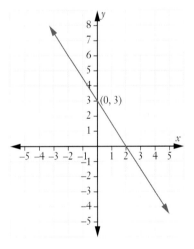

The points $(2, 0)$ and $(-2, 6)$ lie on the line. Confirm that these values for x and y do indeed fit the rule.

Choose another point on the line and similarly confirm that its coordinates fit the rule.

Choose a point that does not lie on the line and confirm that its coordinates do not fit the rule.

Table – rule – graph

If the x and y values are linked by a rule of the form $y = mx + c$ then:

• a straight line, or **linear**, relationship exists between x and y.

• graphing, with the x values on the horizontal axis and the y values on the vertical axis, produces a straight line with gradient m and cutting the vertical axis at the point with coordinates $(0, c)$.

We call c the **vertical intercept**, or y-intercept.

• with the x values increasing by 1, the **common difference** in the y values is equal to m, the gradient of the line.

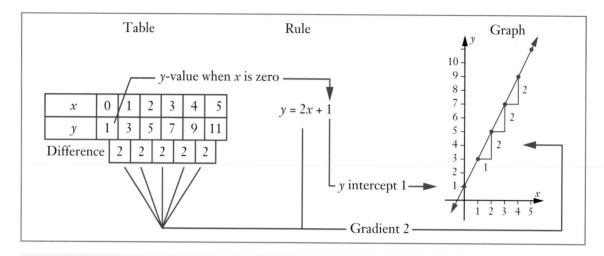

	Table							Rule		Graph

y-value when *x* is zero

x	0	1	2	3	4	5
y	1	3	5	7	9	11
Difference		2	2	2	2	2

$y = 2x + 1$

y intercept 1

Gradient 2

EXAMPLE 1

For each of the following tables determine whether the relationship between the two variables is linear and, for any that are, determine the rule.

a

x	0	1	2	3	4	5
y	−2	1	4	7	10	13

b

P	3	4	5	6	7	8
t	8	15	24	35	48	63

c

r	2	4	1	5	3	6
s	17	31	10	38	24	45

Solution

a

x	0	1	2	3	4	5
y	−2	1	4	7	10	13
Difference		3	3	3	3	3

Constant difference pattern
thus relationship is linear.
The rule will be of the form
$$y = 3x + c.$$
To fit the tabulated data the rule must be
$$y = 3x - 2.$$

b

P	3	4	5	6	7	8
t	8	15	24	35	48	63
Difference		7	9	11	13	15

Difference pattern is not constant
thus relationship is not linear.

c   First present *r* values in order:

r	1	2	3	4	5	6
s	10	17	24	31	38	45
Difference		7	7	7	7	7

Constant difference pattern
thus relationship is linear.
The rule will be of the form
$$s = 7r + c.$$
To fit the tabulated data the rule must be
$$s = 7r + 3.$$

ISBN 9780170390262

Lines parallel to the axes

I   **Lines parallel to the *x*-axis.**

Consider the line parallel to the *x*-axis and passing through $(-2, 3)$, $(1, 3)$ and $(4, 3)$ as shown on the right.

The rule for this line is $y = 3$ because the *y* coordinates of all points lying on this line will equal 3.

Note that this is consistent with the $y = mx + c$ idea because the line has a gradient of zero and a *y* intercept of 3.

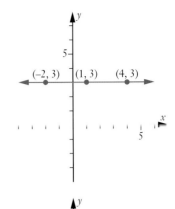

The graph shown on the right shows the horizontal lines:

$$y = 6,$$

$$y = 3,$$

$$\text{and} \quad y = -2.$$

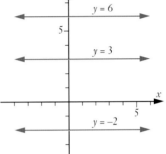

II   **Lines parallel to the *y*-axis.**

The diagram on the right shows some lines drawn parallel to the *y*-axis.

The gradient of each of these vertical lines is undefined because we cannot find the vertical rise in the line for each horizontal unit increased because the line rises vertically for zero increase horizontally! Hence we should not expect the rules for vertical lines to be of the form $y = mx + c$ because the gradient, *m*, is undefined. Indeed straight lines parallel to the *y*-axis are the only straight lines having rules that are *not* of the form $y = mx + c$.

Lines parallel to the *y*-axis have rules of the form $x = c$.

The graph on the right shows the vertical lines:

$$x = -2$$

$$x = 3,$$

$$\text{and} \quad x = 5.$$

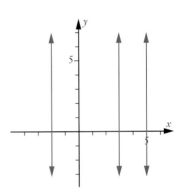

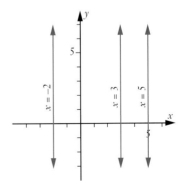

Even though these vertical lines have rules of a different form, points lying on each line must still 'obey' the rule. For example, for a point to lie on $x = 3$ the point must have an *x*-coordinate equal to 3.

Use of a calculator with a graphing facility

Calculators with a graphing facility can display graphs of lines given the rules for the lines.
For example, entering the following rules into such a calculator

$$y = 2x - 3, \qquad y = 2x - 1, \qquad y = 2x + 1, \qquad y = 2x + 3 \qquad \text{and} \qquad y = 2x + 5$$

allows the graphs of these lines to be displayed.

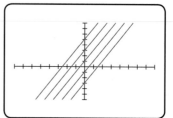

What feature of the five rules allows us to anticipate that the lines would be parallel to each other?

Note: The display on the right shows the line $y = x$.
As we would expect this line has a gradient of 1 and cuts the y-axis at the point $(0, 0)$, i.e. the origin. The line passes through all those points for which the x-coordinate equals the y-coordinate, for example $(0, 0)$, $(1, 1)$, $(2, 2)$ etc.

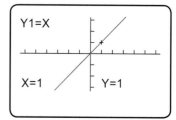

Because the same scale is used on each axis the line $y = x$ makes an angle of 45° with each axis. However, do not expect this 45° property of the line $y = x$ to be evident if different scales are used on each axis. Both of the displays shown below show the line $y = x$ but with different scales used on each axis, and with the two displays using different scales, the two displays appear different and neither shows the 45° nature of the line.

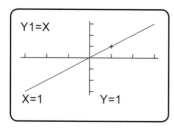

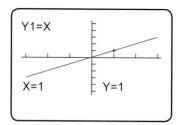

Exercise 8B

For each of questions 1 to 10 determine:

a   the gradient of the line

b   the coordinates of the point where the line cuts the vertical axis

c   the rule for the line.

1

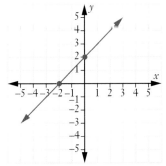

2

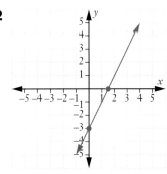

3

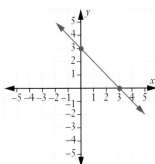

4

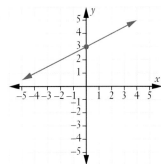

5

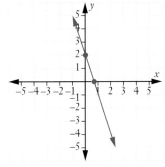

6

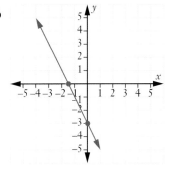

7

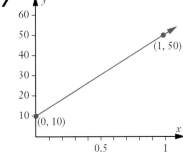

8

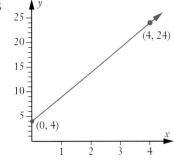

9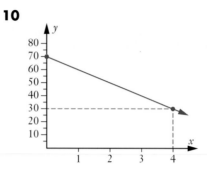

10

11 Determine the rules for each of the lines A to H shown below.

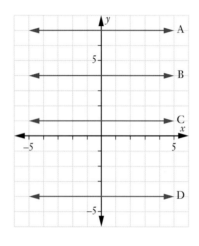

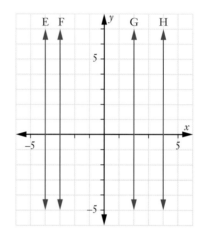

For each of the tables shown in questions 12 to 25, determine whether the relationship between x and y is linear and, for any for which it is, determine the rule for the relationship.

12

x	1	2	3	4	5	6
y	4	7	10	13	16	19

13

x	1	2	3	4	5	6
y	21	17	13	9	5	1

14

x	0	1	2	3	4	5
y	−3	2	7	12	17	22

15

x	0	1	2	3	4	5
y	25	16	9	4	1	0

16

x	1	2	3	4	5	6
y	2	3.5	5	6.5	8	9.5

17

x	5	6	7	8	9	10
y	5	10	20	40	80	160

18

x	−2	−1	0	1	2	3
y	0	1	2	3	4	5

19

x	0	2	4	6	8	10
y	1	3	6	10	15	21

20

x	1	3	5	7	9	11
y	12	10	8	6	4	2

21

x	0	5	10	15	20	25
y	21	31	41	51	61	71

ISBN 9780170390262

22

x	2	5	1	6	4	3
y	10	19	7	22	16	13

23

x	6	3	5	2	1	4
y	18	3	13	−2	−7	8

24

x	5	13	9	3	11	7
y	8	24	16	4	20	12

25

x	16	7	1	13	4	10
y	49	22	4	40	13	31

For Questions 26 and 27 find

a the gradient of the line,

b the coordinates of the point where the line will cut the *y*-axis,

c the rule for the line.

26

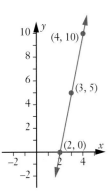

27

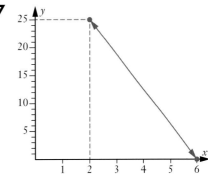

In each of questions 28 to 33 you are given a table of values OR a rule OR a graph.
Use the one you are given to complete the other two. (When drawing the graph assume all *x* values are possible, not just the integer values given in the table.)

28 Table:

x	−4	−3	−2	−1	0	1	2	3	4
y									

Rule: *y* =
Graph:

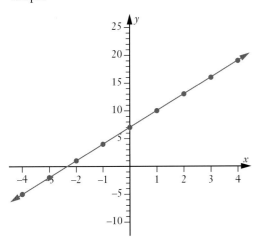

29 Table:

x	−4	−3	−2	−1	0	1	2	3	4
y									

Rule: *y* =
Graph:

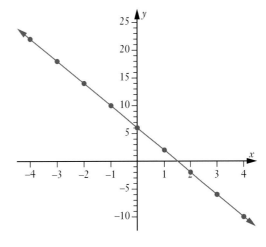

8. Linear relationships ●●●●●●●●●○○○○○○

30 Table:

t	–4	–3	–2	–1	0	1	2	3	4
r	–2	0	2	4	6	8	10	12	14

Rule: $r =$

Graph:

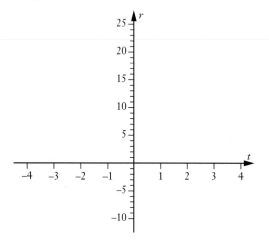

31 Table:

u	–4	–3	–2	–1	0	1	2	3	4
w	–9	–6	–3	0	3	6	9	12	15

Rule: $w =$

Graph:

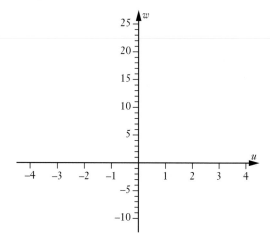

32 Table:

t	–4	–3	–2	–1	0	1	2	3	4
K									

Rule: $K = 4t + 9$

Graph:

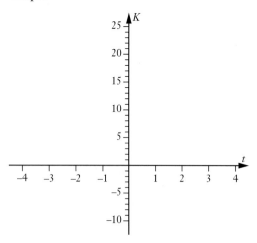

33 Table:

n	–4	–3	–2	–1	0	1	2	3	4
P									

Rule: $P = -2n + 7$

Graph:

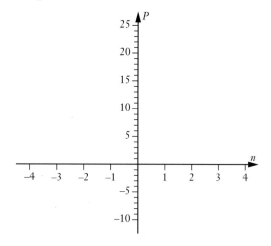

ISBN 9780170390262

34 On squared paper, and with each axis from –6 to 6, use gradients and intercepts to sketch the following four lines on the one graph:

$$y = 2x + 3 \qquad y = -2x + 1 \qquad y = -0.5x + 4 \qquad y = 0.5x - 4$$

Now use your graphic calculator to check the correctness of your sketch.

35 The display on the right shows the lines

$y = 3x$ $\qquad\qquad y = 4x$

$y = 3x - 2$ $\qquad\qquad y = 2.5x + 4$

$y = -1.5x + 3$ $\qquad\qquad y = -\dfrac{2}{3}x + 2$

labelled A to F.

Allocate the correct rule to each line.

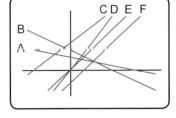

36 The display on the right shows the lines

$y = 0.5x + 2$ $\qquad\qquad y = 0.5x - 2$

$y = -2$ $\qquad\qquad y = -x + 2$

$y = 2x - 2$

labelled A to E.

Allocate the correct rule to each line.

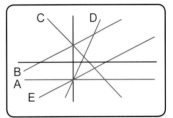

37 The display on the right shows the lines

$y = 30x$ $\qquad\qquad y = 30$

$y = 30x - 90$ $\qquad\qquad y = 10x + 60$

$y = -10x + 30$

labelled A to E.

Allocate the correct rule to each line.

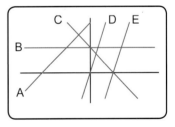

38 Both of the displays shown below show the line $y = 2x + 3$. How is it that they can look different? Explain.

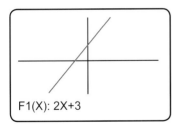

F1(X): 2X+3

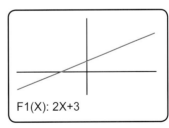

F1(X): 2X+3

Linear graphs

More about $y = mx + c$, the equation of a straight line

From earlier work we know that straight lines have equations that are of the form:

$$y = mx + c$$

1 Why should equations of the form $y = mx + c$ produce straight line graphs?

A straight line must clearly have a constant gradient. Thus wherever we are on the line, if we journey one further unit horizontally, then the line will always rise a constant vertical amount.

Consider the equation $y = mx + c$.

- If x changes from 2 to 3, y changes from $(2m + c)$ to $(3m + c)$, a rise of m.
- If x changes from 3 to 4, y changes from $(3m + c)$ to $(4m + c)$, a rise of m.
- If x changes from 98 to 99, y changes from $(98m + c)$ to $(99m + c)$, a rise of m.

Each unit increase in the x-coordinate produces an increase of m units in the y-coordinate. i.e. a straight line of gradient m.

If an equation cannot be expressed in the form $y = mx + c$ then the y-coordinate will not rise by a constant amount for each unit increase in the x-coordinate.

For example, consider $y = x^2$.

- If x changes from 2 to 3, y changes from 4 to 9, i.e. a rise of 5.
- If x changes from 3 to 4, y changes from 9 to 16, i.e. a rise of 7.
- If x changes from 98 to 99, y changes from 9604 to 9801, i.e. a rise of 197.

Each increase of one unit in the x-coordinate does not produce the same change in the y-coordinate. i.e. not a straight line graph.

2 The equation of a line is like the membership ticket for the line.

As was mentioned earlier, if a point lies on a particular line then the coordinates of that point will 'fit' the equation of the line, and if it does not lie on the line the coordinates will not 'fit' the equation. The equation of a line is the rule which all points lying on the line must 'obey'. In this way the equation is like the membership ticket for the line. If a point does not fit the equation it cannot lie on the line.

EXAMPLE 2

State whether each of the points A to C lie on the line $y = 2x + 7$.

$$A\,(1, 8) \qquad B\,(3, 13) \qquad C\,(-5, -3)$$

Solution

If A $(1, 8)$ lies on the line then $x = 1$ and $y = 8$ must 'fit' the equation.
Substituting $x = 1$ into $y = 2x + 7$ gives $y = 9$.
Thus point A $(1, 8)$ *does not* lie on the line $y = 2x + 7$.

If B $(3, 13)$ lies on the line then $x = 3$ and $y = 13$ must 'fit' the equation.
Substituting $x = 3$ into $y = 2x + 7$ gives $y = 13$.
Thus point B $(3, 13)$ *does* lie on the line $y = 2x + 7$.

If C $(-5, -3)$ lies on the line then $x = -5$ and $y = -3$ must 'fit' the equation.
Substituting $x = -5$ into $y = 2x + 7$ gives $y = -3$.
Thus point C $(-5, -3)$ *does* lie on the line $y = 2x + 7$.

ISBN 9780170390262

It may not look like $y = mx + c$ but it may still be linear

Straight lines have equations that can be written in the form $y = mx + c$.

Consider each of the following equations:

$$2y + 3x = 12$$
$$5x - 2y = 15$$
$$\frac{y}{3} = \frac{1}{2} - \frac{2x}{5}$$

Each of these can be rearranged into the form $y = mx + c$:

$$2y + 3x = 12$$
$$2y = -3x + 12$$
$$y = -1.5x + 6$$

$$5x - 2y = 15$$
$$5x - 15 = 2y$$
$$y = 2.5x - 7.5$$

$$\frac{y}{3} = \frac{1}{2} - \frac{2x}{5}$$
$$\times \text{ by } 3 \quad y = -\frac{6x}{5} + \frac{3}{2}$$

Hence whilst they may not initially look like $y = mx + c$ each of the three equations are equations of straight lines.

Determining the equation of a straight line

We can determine the equation if we know:

- the gradient and the vertical intercept (see example 3 that follows)
- or • the gradient and one point on the line (see example 4)
- or • if we know two points on the line (see example 5).

EXAMPLE 3

(Given the gradient and the vertical intercept)
State the equation of the straight line that cuts the y-axis at the point $(0, 1)$ and has a gradient of 6.

Solution

A line with gradient m and cutting the y-axis at $(0, c)$ has equation $y = mx + c$.
Thus the given line has equation $y = 6x + 1$.

EXAMPLE 4

(Given the gradient and one point on the line)
Find the equation of the straight line through the point $(4, -3)$ and with a gradient of -2.

Solution

A straight line of gradient m has an equation of the form $\quad y = mx + c$.
Thus the given line will have an equation of the form $\quad y = -2x + c$.
The line passes through the point $(4, -3)$.
Thus the values $x = 4$ and $y = -3$ must 'fit' the equation, i.e. $\quad (-3) = -2(4) + c$
giving $\quad c = 5$.

Thus the given line has equation $y = -2x + 5$.

---

8. Linear relationships ●●●●●●●●○○○○

EXAMPLE 5

(Given two points that lie on the line)
Find the equation of the straight line through the points (–2, 8) and (4, –1).

Solution

Starting at the point with the lower x-coordinate (–2, 8), and moving to the point (4, –1), we travel across 6 units and down 9 units. Thus in moving across 1 unit we travel *down* $\frac{9}{6}$ units, i.e. $\frac{3}{2}$ units. The gradient of the line is $-\frac{3}{2}$.

Thus the given line will have an equation of the form $\qquad\qquad y = -1.5x + c.$

The line passes through the point (4, –1).
Thus the values $x = 4$ and $y = -1$ must 'fit' the equation, i.e. $\qquad -1 = -1.5(4) + c$
$\qquad\qquad\qquad\qquad\qquad\qquad\qquad\qquad\qquad$ giving $\qquad c = 5.$

Thus the given line has equation $y = -1.5x + 5.$

(The reader should confirm that using the point (–2, 8) and saying that the values $x = -2$ and $y = 8$ must 'fit' the equation also gives $c = 5$.)

A useful rule

Finding the gradient between two points on a line

A useful rule to remember when determining the gradient of the line through two points, A and B is:

$$\text{Gradient} = \frac{\text{the change in the } y\text{-coordinate in going from A to B}}{\text{the change in the } x\text{-coordinate in going from A to B}}.$$

Thus if A has coordinates (x_1, y_1) and B has coordinates (x_2, y_2) then the gradient of the line through A and B $= \dfrac{y_2 - y_1}{x_2 - x_1}$.

Note: In this formula $\dfrac{y_1 - y_2}{x_1 - x_2}$ would also give the correct answer but $\dfrac{y_1 - y_2}{x_2 - x_1}$ and $\dfrac{y_2 - y_1}{x_1 - x_2}$ would not. Hence make sure that 'whichever point you get the first y-coordinate from is also where you get the first x-coordinate from'.

Calculator routines

Your calculator may have programmed routines that allow the equation of a line to be determined simply by inputting the coordinates of two points on the line, or inputting the gradient and the coordinates of just one point on the line. Such routines can be useful but make sure that you understand the underlying ideas and can apply them without the assistance of calculator programs if required.

Exercise 8C

1 Calculate the gradient of the straight line through each of the following pairs of points.

   a   (4, 6) and (2, 2)   **b**   (6, 7) and (5, 3)   **c**   (4, 5) and (2, 1)

   d   (6, 7) and (2, 5)   **e**   (5, 3) and (1, 2)   **f**   (5, 3) and (4, 2)

   g   (4, 3) and (2, 7)   **h**   (5, 2) and (3, –3)   **i**   (4, 2) and (–2, –1)

   j   (1, –7) and (–1, 1)   **k**   (–1, –2) and (1, 3)   **l**   (2, –3) and (6, –1)

2 State the gradient and the coordinates of the y axis intercept of each of the following straight lines.

   a $y = 3x - 17$ **b** $y = -2x + 13$ **c** $y = 5 - 7x$

   d $2x + 3y = 24$ **e** $5y + 2x = 8$ **f** $2x - 3y + 9 = 0$

   g $\dfrac{x}{2} + y = 11$ **h** $\dfrac{y}{5} + \dfrac{x}{2} = 3$ **i** $\dfrac{2x}{5} + \dfrac{y}{3} = 4$

3 What is the equation of the x axis?

4 What is the equation of the y axis?

5 State whether each of the points A to E lie on the line $y = 3x - 5$.

   A (6, 12)   B (5, 11)   C (2, 1)   D (–3, –13)   E (–1, –8)

6 State which of the points F to J do *not* lie on the line $y = -x + 6$.

   F (1, 5)   G (0, 6)   H (2, 8)   I (–1, 4)   J (6, 0)

7 Write down the equation of the straight line with gradient 3 and cutting the y-axis at (0, 4). Does this line pass through the point (–1, 1)?

8 Write down the equation of the straight line cutting the y-axis at (0, 2) and with gradient 0.5. Which of the following points lie on this line?

   A (2, 1)   B (2, 0)   C (4, 2)   D (–6, –1)   E (4, 4)

9 Given that all of the points A to F given below lie on the line $y = 2x - 5$ determine the values of a, b, c, d, e and f.

   A (3, a)   B (2, b)   C (–4, c)   D (2·5, d)   E (e, 13)   F (f, –5)

10 Find the equation of each of the following straight lines.

   a   Gradient 1, through (3, 5).   **b**   Gradient –1, through (6, –1).

   c   Gradient –2, through (3, 2).   **d**   Gradient 5, through (–2, –2).

   e   Gradient $\dfrac{1}{2}$, through (8, 9).   **f**   Gradient $-\dfrac{1}{2}$, through (–3, 0).

   g   Gradient $\dfrac{3}{2}$, through (9, 2).   **h**   Gradient $-\dfrac{1}{3}$, through (7, –1).

11 Find the equation of each of the following straight lines.

   a   Through (2, 5) and (6, 9).   **b**   Through (0, –1) and (2, –9).

   c   Through (14, 1) and (16, –5).   **d**   Through (1, 1) and (2, 3).

   e   Through (1, 2) and (13, 6).   **f**   Through (3, –2) and (–1, 6).

   g   Through (3, 9) and (0, 4).   **h**   Through (0, 5) and (2, –5).

Linear relationships in practical situations

Situation One at the start of this chapter involved copies of a particular book. Each copy weighed 1.5 kg so one book would weigh 1.5 kg, two books 3 kg, three books 4.5 kg, four books 6 kg and so on.

In this example the relationship between the total weight, w kg, and the number of copies, c, is exactly linear with all points exactly lying on the straight line

$$w = 1.5c$$

This is an example of a linear relationship in real life.

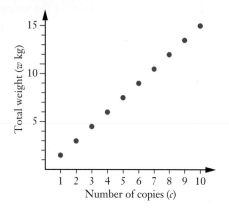

Some other situations may involve data that does not exactly fit a straight line but for which a straight line may be a reasonable *model* to use to summarise what is going on, and that might allow reasonable predictions and statements to be made.

Suppose, for example, that in the process of monitoring the survival of a particular endangered species of animal the numbers of these animals held in zoos around the world is recorded every two years over a period of twenty years. Suppose the numbers were as shown in the graph below left. Whilst the values do not exactly lie in a straight line, the rule $N = 28 + 1.5t$ could be a reasonable linear model to use for this data, as shown below right.

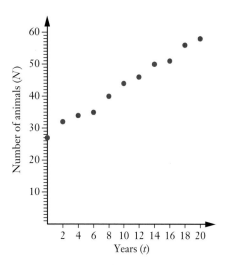
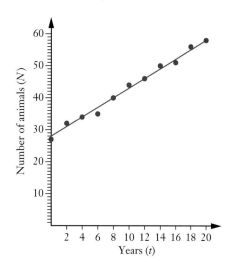

In this linear model ($N = 1.5t + 28$):

- the vertical intercept is 28, indicating that the number of these animals kept in zoos at the beginning of the 20 year period was approximately 28,

- the gradient is 1.5, indicating that on average each 1 year increase saw an increase of 1.5 in the number of these animals kept in zoos (i.e. an increase of approximately 3 every two years).

ISBN 9780170390262

Exercise 8D

1 If we plot degrees Celsius, (°C), on the x-axis and degrees Fahrenheit, (°F), on the y-axis, the graph for converting from one scale to the other is a straight line.

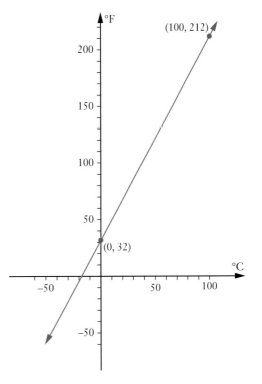

   a Given that 100°C is the same as 212°F and 0°C is the same as 32°F find the equation of the line in the form

$$F = mC + b,$$

   where m and b are constants.

   b What does the value of m tell us about °C and °F temperatures?

   c Convert 55°C to °F.

   d Convert –10°C to °F.

   e Convert 59°F to °C.

   f Convert –4°F to °C.

   g Is there a temperature for which the number of degrees Celsius is the same as the number of degrees Fahrenheit, and if so what is that temperature?

2 When a particular spring has a mass of M kg suspended from one end the total length of the spring is L metres where

$$L = kM + L_0 \text{ where } k \text{ and } L_0 \text{ are constants.}$$

   a What will the value of k tell us about this situation?

   b What will the value of L_0 tell us about this situation?

   c A graph of M plotted on the x-axis and L on the y-axis passes through the points $(2, 0.85)$ and $(3, 1.05)$.

   Calculate k and L_0 and hence determine how much the spring is extended **beyond its natural length** when a mass of 250 g is suspended from it.

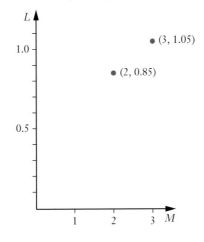

3 The diagram on the right shows the proposed layout of a small airfield. The diagram shows the main runway, the approach lights, the warning lights and the administration building.

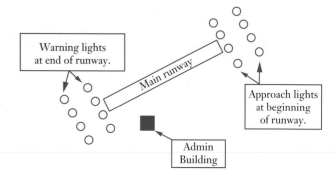

The second diagram shows the proposal as a graph with lengths in metres and the admin building as the origin.

Find:

a the coordinates of the points A, B, C, D, E and F, (all divisible by 20)

b the equation of the straight line through A and B

c the equation of the straight line through C and D

d the equation of the straight line through E and F.

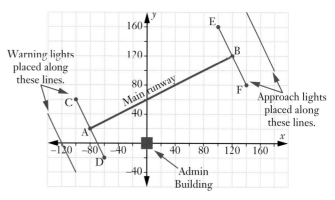

4 If we plot the 'Number of metered units', N, on the x-axis and the 'Amount to be paid', A, on the y-axis then the graph for calculating a telephone bill from one particular company is a straight line with equation

$$A = mN + c,$$

where m and c are constants.

In the context of this question:

a what does the value of m tell us about A and N?

b what does the value of c tell us?

c If the bill for 100 units is $64 and for 150 units is $76, determine the equation of the line.

d What would be the bill for 200 units?

e If the bill was for $82 how many units were used?

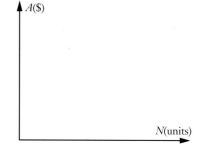

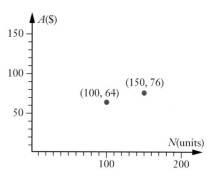

ISBN 9780170390262

5 The diagram below left shows the proposed road system for a new housing estate off an existing road 'Baxton Drive'. The computer models this system graphically as shown below right, with 1 unit on each axis representing 25 metres.

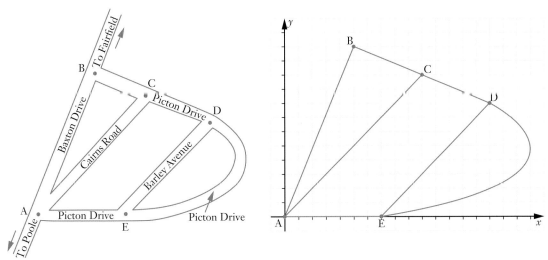

Find the equation of the straight line through

a A and B,   **b** A and C,   **c** A and E,

d E and D,   **e** B, C and D.

6 Susie Fuse, an electrician, charges her customers a standard call out fee plus a certain amount per hour. With this method of charging the cost to the customer, $C, and the time taken to do the job, T hours, are linearly related and follow a rule of the form

$$C = mT + c.$$

Explain what information m, the gradient, and c, the intercept with the vertical axis, are giving in this context.

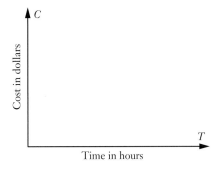

For a job that takes her three hours Susie charges $440 and for a job that takes her four and a half hours hours she charges $620.

Write the rule $C = mT + c$ with m and c evaluated.

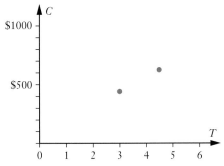

7 A taxi company charges a fixed start fee (called the 'flag fall' in the taxi industry) of $4.90 followed by a charge of $1.85 per kilometre.

If the cost for a journey of x kilometres is $$C$, write a rule in the form

$$C = c + mx.$$

8 A water tank initially contains 1000 litres of water.

The tank develops a leak at its base, which causes water to leak out at a constant rate of 200 millilitres every minute.

If V litres is the volume of water in the tank t minutes after the leak commenced, write a rule relating V and t.

9 It costs a small business $800 to run the business each week.

The company imports just one type of product from overseas and sells each one for $75 more than it buys each one for.

Find an expression for $$P$, the weekly profit the company makes, if it sells n of these products in that week.

10 A company sells copies of a book that it prints on a book by book basis, as orders arrive. Each book costs the company $12 to produce in this way.

The company sells each book for $22, which includes 10% goods and services tax.

The company forwards the goods and services tax amount to the government and the company keeps the rest.

If the company makes a profit of $$P$ when it sells x copies of this book write a rule for P in the form $P = \mathrm{m}x$.

11 To cook a joint of meat a recipe book advises preheating the oven to 180°C and then, when the oven temperature has reached this temperature, place the meat in the oven for '20 minutes per kilogram + 20 minutes over'.

Express as a rule the time, t hours, a joint of meat weighing k kilograms should be placed in the hot oven for it to cook according to these instructions.

12 A linear relationship exists between the profit, $$P$, that the organisers of a concert make, and N, the number of tickets they sell. With P plotted on the vertical, y, axis and N on the horizontal, x, axis the line of this relationship passes through the points (900, 400) and (1100, 1300).

Find the equation of this line in the form

$$P = mN + c,\text{ where } m \text{ and } c \text{ are constants.}$$

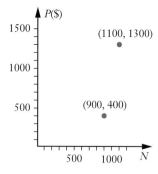

a What does the value of m tell us in this context?

b What does the value of c tell us in this context?

c What will be the profit when 1500 tickets are sold?

d If the concert hall has a maximum capacity of 2500 what profit will the organisers make if they give away 150 complimentary tickets and sell all the rest?

e What is the least number of tickets the organisers could sell and still not make a loss?

ISBN 9780170390262

13 The owner of a computer shop calculates that his weekly profit from computer sales is linearly related to the number of computers sold that week.

If he sells 10 computers in a week his total profit is $560.

If he only sells 5 computers in the week he makes a profit of $10.

The rule relating his total profit for the week, P, to the number of computers sold, x, is given by:

$$\text{Total profit in dollars } (P) = mx - c,$$

c being the fixed weekly cost of running the shop.

a Calculate m and c.

b What is his weekly profit from computer sales in a week that he sells 20 computers?

14 The membership secretary of a club monitors the growth in membership over a 5 year period. Plotting the 5 years (1, 2, 3, 4 and 5) on the horizontal 't' axis and the membership numbers on the vertical, 'N', axis the secretary finds there is an almost perfect linear relationship between t and N. Express the relationship in the form $N = mt + c$ given that when $t = 1, N = 250$ and when $t = 5, N = 410$.

What do the values of the gradient, m, and the vertical intercept, c, tell us in the context of this question?

Use your equation to predict the value of N when $t = 10$, assuming the linear relationship continues.

15 The table below shows the profit, P, that a company makes from the sale of x copies of a particular book it has had printed.

x	0	100	200	300	400	500	600	700	800
P	−3750	−2250	−750	750	2250	3750	5250	6750	8250

a Determine the rule for the relationship between P and x.

b How many copies of the book must the company sell to achieve a profit of more than $10 000?

16 The monitoring of the numbers of a particular endangered species of animal found that over a number of years, from the time the monitoring started, the numbers thought to be in existence in the wild showed a steady decline. Indeed with N representing the number of these animals thought to be in existence in the wild, t years into the monitoring program, N and t were approximately following the rule:

$$N = 5740 - 350t.$$

a Interpret what the numbers 5740 and 350 mean in the context of this situation.

b Graph the rule $N = 5740 - 350t$ with N plotted on the vertical axis and t on the horizontal axis.

c State the coordinates of the point where the line $N = 5740 - 350t$ cuts the horizontal axis and explain the significance of this point in the context of this question.

Miscellaneous exercise eight

This miscellaneous exercise may include questions involving the work of this chapter, the work of any previous chapters, and the ideas mentioned in the Preliminary work section at the beginning of the book.

1 Determine the equation of each of the straight lines A to I shown below.

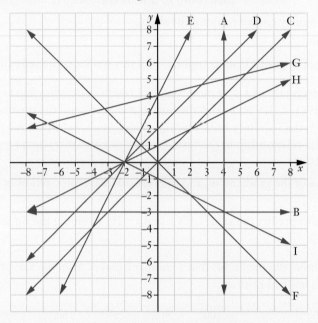

2 A particular straight line with a gradient of b and cutting the y-axis at the point with coordinates $(0, a)$ has equation $y = a + bx$. The line passes through the point $(5, -2)$.

Which of the following equations must be true?

Equation 1	Equation 2	Equation 3
$5 = a - 2b$	$y = 5 - 2x$	$a + 5b + 2 = 0$

3 Find the values of $a, b, c, \ldots h$ in the following.

  a $a : 2 = 6 : 5$ **b** $6 : 7 = 3 : b$ **c** $5 : 2 = 15 : c$ **d** $d : 2 = 19 : 10$

  e $e : 3 = 9 : 2$ **f** $2 : 3 = 9 : f$ **g** $4 : g = 8 : 9$ **h** $4 : h = 9 : 8$

4 The ratio of females to males in a particular workforce is $7 : 9$.

There are 252 males in this workforce.

How many females are there in this workforce?

5 In the first three units of a course a student achieves a mean of 64%. In the next two units the student achieves a mean of 51%. What is the least mark the student needs to achieve in the sixth and final unit to gain an overall mean of at least 60%?

6 Four distributions of marks have both their dot frequency diagrams and their boxplots shown below. Without the assistance of a calculator, match each dot frequency with its corresponding boxplot.

Dot frequency 1:

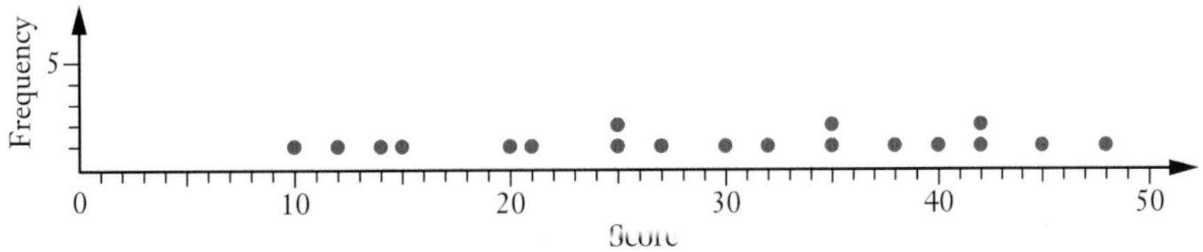

Dot frequency 2:

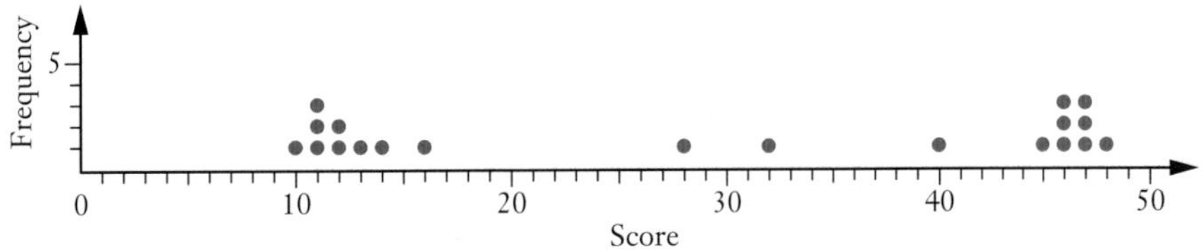

Dot frequency 3:

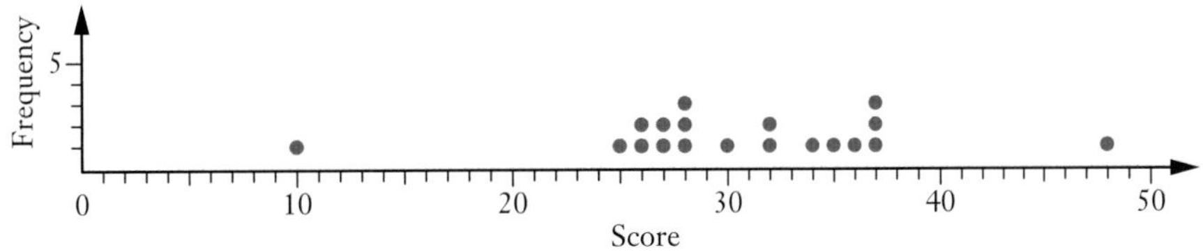

Dot frequency 4:

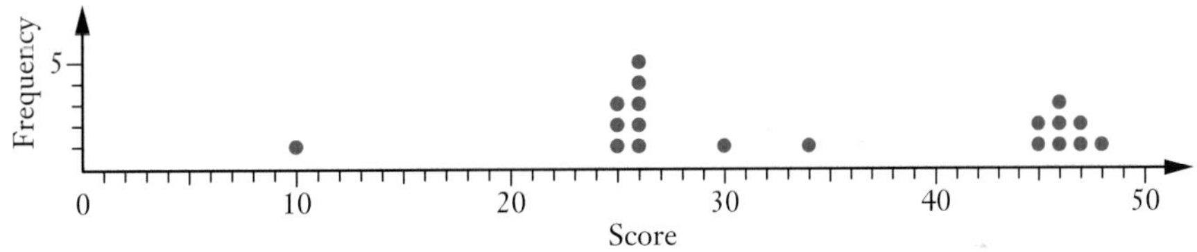

Box plot A:

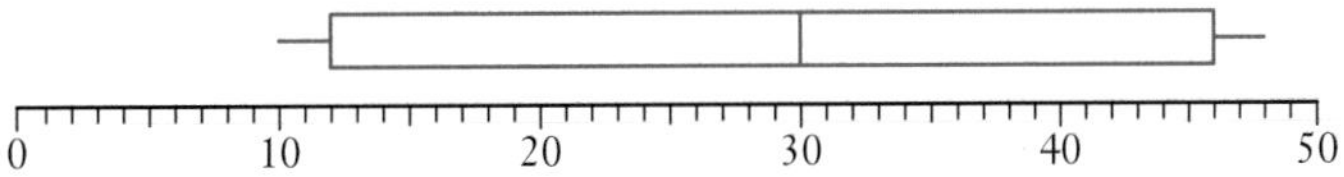

Box plot B:

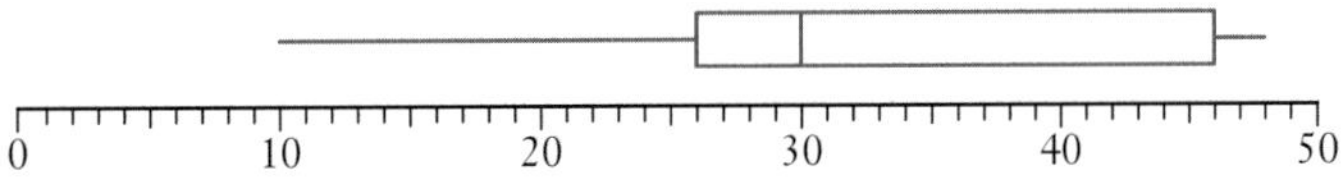

Box plot C:

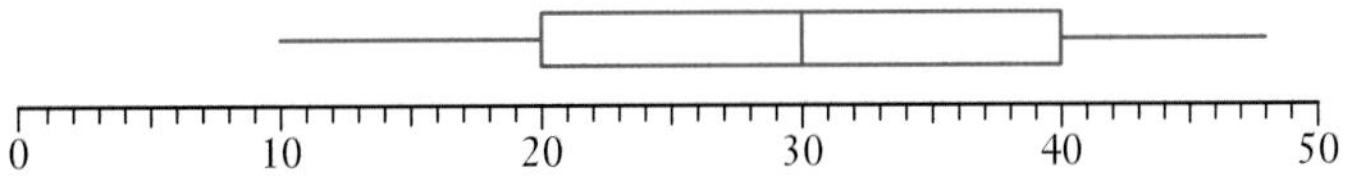

Box plot D:

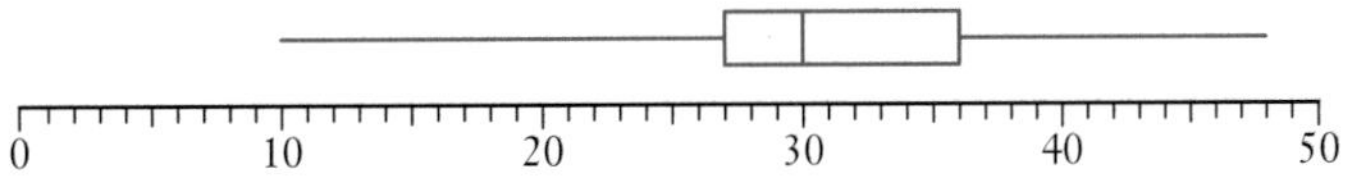

8. Linear relationships ●●●●●●●●○○○○○ 153

7 Forty one students sat a test. The boxplot of their results is shown below.

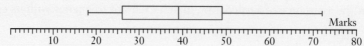

With the student who gained the top mark ranked 1st, the students ranked 19th and 21st in the test scored identical marks.

The top twelve students scored twelve different marks.

The student ranked 11th in the test scored 47.

    a   What was the top mark achieved in the test?

    b   What score did the student ranked 20th achieve?

    c   What score did the student ranked 10th achieve?

8 The three 'octapatterns' below all follow the pattern shown on the right.

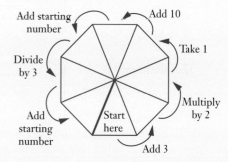

Copy and complete each of the octapatterns shown below.

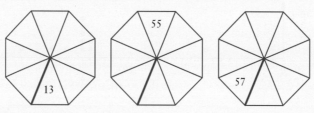

9 Certain types of cricket chirp more frequently as the temperature rises. For a particular species the number of chirps per minute (N) is found to be related to the Celsius temperature ($C°$) according to the rule:

$$N \approx 7C - 16$$

    a   If we were to graph $N = 7C - 16$, with C on the horizontal axis and N on the vertical axis it would give a straight line with gradient 7 and intersecting the vertical axis at -16. Interpret these two numbers in the context of this 'chirping crickets' situation.

    b   What does the rule suggest as the temperature below which we would not expect a cricket to chirp?

    c   Roughly how many chirps will a cricket of this species make per minute if the temperature is

        i 14°C?   **ii** 28°C?

    d   Estimate the temperature if a cricket of this species is chirping

        i 200 times per minute   **ii** 50 times per 20 seconds.

9.

Piecewise defined relationships

- Piecewise defined relationships
- Miscellaneous exercise nine

For each of the three situations given below choose the one graph from the six shown that best fits the situation. Then, for each situation, having chosen the most appropriate graph, make a sketch of the graph and include labels and numbers on each axis. (If you think that none of the graphs fit the situation draw your own appropriate graph.)

Situation One

The income tax system in Australia is what is known as a 'progressive system'. This means that the *rate* of income tax increases as a person's taxable income increases.

For this situation suppose that the following progressive system were to apply:

Taxable income	Rate at which tax is deducted
$0 to $20 000	Nil
$20 001 to $50 000	20% of every dollar over $20 000
$50 001 to $80 000	$6 000 plus 40% of every dollar over $50 000
Over $80 000	$18 000 plus 60% of every dollar over $80 000

Situation Two

To deliver a parcel from town A to town B a company charges $7.50 for the first kilogram, or part thereof, and then a further $2.50 per kilogram, or part thereof, after that. Thus a parcel weighing 0.56 kg will cost $7.50, a parcel weighing 2.4 kilograms will cost $12.50 (= $7.50 + 2 × $2.50), a parcel weighing 5.1 kilograms will cost $20 (= $7.50 + 5 × $2.50) etc.

Situation Three

John has bought a new racing bike and has sold his old one to Peter. Peter lives 12 kilometres from John's house, along an almost straight road. John rides the bike to Peter's house, stays there for 30 minutes having a chat with Peter, and then walks back to his own house. He cycles at a steady 12 km per hour and walks at a steady 6 km per hour.

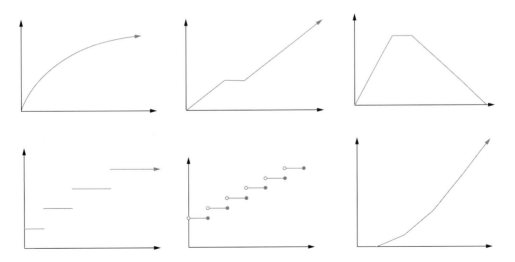

Piecewise defined relationships

The three situations on the previous page (and five of the six graphs) involved circumstances in which the relationship between two variables involved a number of different linear relationships. Which relationship applied depended on where on the horizontal axis we were. In this way different rules applied for different *pieces* of the horizontal axis. Such relationships are said to be **piecewise defined**.

One of the six graphs on the previous page is not a piecewise defined linear graph – which one?

Two of the six graphs on the previous page are also called **step graphs** – which two?

Consider, for example, the graph on the right.

In this case, for x less than -4 one rule applies, for x from -4 to -1 another rule applies, etc.

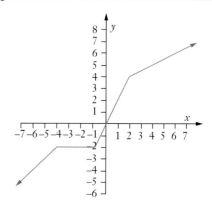

We write this as:

$$\text{For } x < -4, \qquad y = x + 2$$
$$\text{for } -4 \le x < -1, \qquad y = -2$$
$$\text{for } -1 \le x < 2, \qquad y = 2x$$
$$\text{for } x \ge 2, \qquad y = 0.5x + 3$$

(Note that whilst in the above listing we have attached the $x = -4$ value to the $y = -2$ rule it could equally well have been attached to the $y = x + 2$ rule. Similarly the $x = -1$ and $x = 2$ values could be differently 'attached'.)

In the piecewise defined relationship shown on the right the filled circle shows where the value for $x = 2$ **is**, and the open circle shows where the value for $x = 2$ **is not**. (Did you notice this aspect in one of the situations on the previous page?)

In this case

$$\text{For } x < -3, \qquad y = -3$$
$$\text{for } -3 \le x \le 2, \qquad y = 2x + 3$$
$$\text{for } x > 2, \qquad y = -x + 6$$

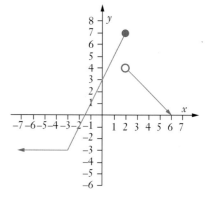

The above examples involved abstract lines without any real context attached but, as the first three situations demonstrated, piecewise defined relationships do occur in real life.

For example consider the situation of a company paying commission for sales achieved by its sales people according to the following rules:

Sales	Commission
$0 \to $1000	5% of sales
$1000 \to $2000	$50 + 10% of each $1 over $1000
$2000 and over	$150 + 15% of each $1 over $2000

The graph of this situation is shown on the right.

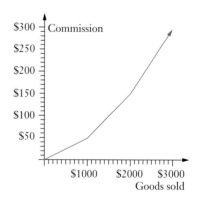

Exercise 9A

1 Copy and complete the following statements for the piecewise defined function shown on the right.

For $x < 0$, $y =$

for $0 \le x < 4$, $y =$

for $x \ge 4$, $y =$

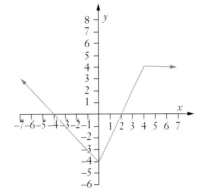

2 Copy and complete the following statements for the piecewise defined function shown on the right.

For $x \le -4$ $y =$

for $y =$

for $y =$

for $y =$

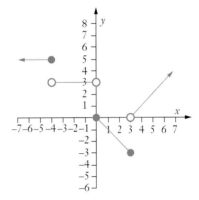

3 Draw the following piecewise defined function:

For $x \le -5$, $y = -x - 5$

for $-5 < x \le 3$, $y = x + 5$

for $x > 3$, $y = 8$.

4 Draw the following piecewise defined function:

For $x \le 0$, $y = -3$

for $0 < x < 4$, $y = 2x$

for $x = 4$, $y = 10$

for $x > 4$, $y = -x + 7$.

5 The fare charged by a taxi company depended upon the number of minutes the journey lasted. The graph shows the charge in dollars graphed against the time of the journey in minutes. Write a few sentences describing your interpretation of the situation based on information given by the graph.

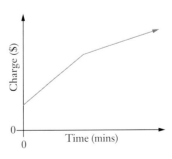

6 The distance time graph on the right is for two brothers journeying from home to the same school. One brother leaves early and walks to school and the other leaves later and cycles.

    a Does the blue broken line ' – – – – ' represent the journey of the walker or the cyclist?

    b Estimate the time when the cyclist passes the walker.

    c How many minutes did the walker take to walk to school?

    d What was the steady speed maintained by the walker during his walk?

    e How many minutes did the cyclist take to ride to school?

    f What was the steady speed maintained by the cyclist during his ride?

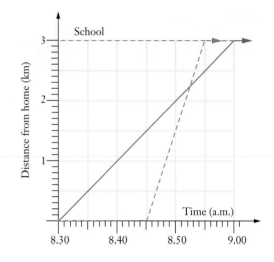

7 The distance time graph shown on the right is for the motion of a cyclist travelling from town A to town B, 60 km away, and a delivery truck making the round trip from A to B and back to A again.

    a When did the cyclist leave town A?

    b When did the cyclist reach town B?

    c The cyclist stopped twice for a rest. How long was each stop?

    d What speed did the cyclist maintain

        i prior to the first stop?

        ii between the two stops?

        iii after the second stop?

    e What speed did the delivery truck maintain

        i from town A to B?

        ii from town B back to A?

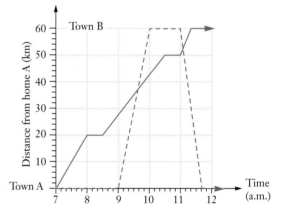

    f Estimate the time and distance from A of the place where the delivery truck passed the cyclist when they were both travelling towards B.

    g Estimate the time and distance from A of the place where the delivery truck passed the cyclist when the truck was returning to A.

ISBN 9780170390262

8 A long straight road links three towns A, B and C with B between A and C. From town A it is 130 km to B and a further 140 km to C. A truck leaves A at 8 a.m. and travels to B. For the first half hour the truck maintains a steady speed of just 60 km/h due to speed restrictions. After this the truck is able to maintain a higher speed and arrives in town B at 9.30 a.m. Unloading and loading in town B takes 1 hour and then the truck travels on to C maintaining a steady 80 km/h for this part of the journey.

A car leaves A at 9 a.m. that same morning and travels directly to C. Subject to the same speed restrictions it too maintains a steady 60 km/h for the first half hour.

After this first half hour the car then maintains a steady 100 km/h all the way to town C.

Draw a distance time graph for this situation and use your graph to answer the following questions:

a When does each vehicle reach town C?

b What steady speed did the truck maintain from 8.30 a.m. to 9.30 a.m.?

c What was the average speed of the truck from A to B? (to nearest km/h.)

d When and where did the car pass the truck?

9 The graph on the right shows a typical income tax system. The higher an individual's taxable income the more tax the person must pay.

Use the graph to estimate the tax payable by someone with a taxable income of

a $30 000   **b** $40 000

c $48 000   **d** $3000

If a person has to pay tax of $20 000 what is their taxable income?

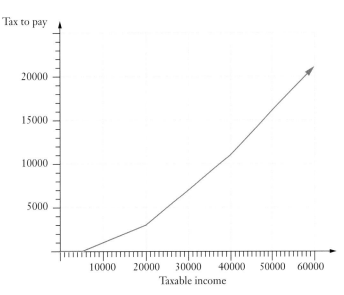

10 Draw the graph for the commission paid by a company to its sales people if payments are made according to the following rules:

Sales	Commission
$0 → $5000	4% of sales
$5000 → $10 000	$200 + 6% of each $1 over $5000
Over $10 000	$500 + 10% of each $1 over $10 000

11 When a real estate agent arranges for the sale of a house the owner of the house pays the agent a fee, often based on the amount the house sells for.

Let us suppose that one agent's fee structure is as shown in the graph below.

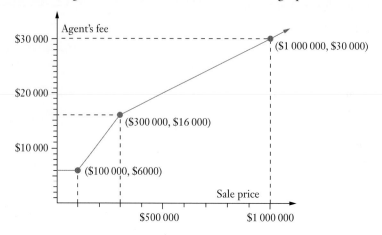

Express this piecewise defined relationship as follows by copying and completing the sentences.

For the first $_____ of the sale price the agent's fee is a fixed $_____.

From $_____ to $_____ the fee is $_____ plus _____% of the amount over $_____.

From $_____ and over the fee is $_____ plus _____% of the amount over $_____.

12 Suppose water is flowing at a constant rate into a container. For a container shaped as shown on the right the graph of the height of the water level plotted against time would be as shown.

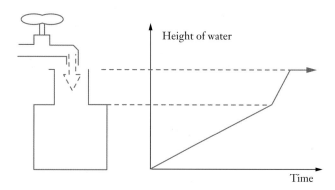

Sketch the graph of the height of the water level against time for each of the following containers.

a   **b** 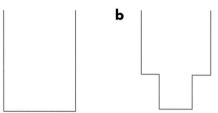   **c**    **d**

ISBN 9780170390262

Miscellaneous exercise nine

This miscellaneous exercise may include questions involving the work of this chapter, the work of any previous chapters, and the ideas mentioned in the Preliminary work section at the beginning of the book.

1 Find the rule that exists between P and t given the following table.

t	2	3	4	5	6	7
P	1	4	7	10	13	16

2 Given that the relationship between x and y is linear find the values of $a, b, c, \ldots g$.

x	0	1	2	3	4	5	6	...	f	g
y	a	b	c	14	d	24	e	...	54	494

3 Write the equations of each of the lines A to J shown in the graphs below.

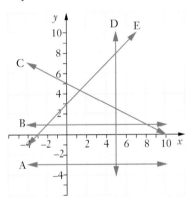 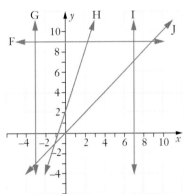

4 A company determines that the cost, $\$C$, for the production of x radios of a particular type is given by: $C = 5200 + 16x$

  a Interpret the 5200 and 16 in this equation in the context of this question.

  Find the mean cost per radio when

  b 100 radios are produced,

  c 500 radios are produced,

  d 1000 radios are produced.

5 The ratio of year eight students in a school to non year eight students in the school is 7 : 25. If there are 960 students in the school altogether how many year eight students are there in the school?

6 What number gives you the same answer when you add sixteen to it as when you multiply it by 5?

7 I think of a number, double it, add five, multiply the result by four, take away the number I first thought of and end up with sixty two. Find the number first thought of.

8 A set of numbers consists of 2 fours, 8 fives, 11 sixes, 9 sevens and a number of eights. If the mean of the set is 6.2 determine the number of eights in the set.

9 The histogram and the box plot for a data set are shown below.

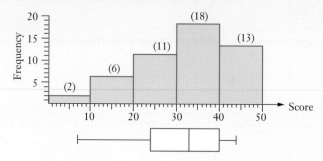

Determine

a the median,

b the range,

c the interquartile range,

d the modal class.

Use your calculator to determine an estimate for the mean.

10 Two sets of students sat the same test and the boxplots of their marks are shown below.

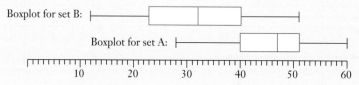

a What percentage of students in set A exceeded the highest mark obtained by students in set B?

b Which two features of the boxplots suggest that the marks in set B were more variable (i.e. more spread out) than those in set A?

If the top 25% of the students in set B, as defined by the test results, were all moved to set A how would:

c the median mark of those left in set B compare with that of set B before the move?

d the range of the scores of those left in set B compare with that of set B before the move?

e the range of scores of the new set A compare with the range of the scores of set A before the move?

f the interquartile range of the new set A compare with the interquartile range of the scores of set A before the move?

ISBN 9780170390262

10.

Trigonometry for right triangles

- Right angled triangles
- Trigonometry
- Hypotenuse, opposite and adjacent
- Notes regarding calculator usage
- Applications
- Accuracy and trigonometry questions
- Bearings
- Elevation and depression
- More vocabulary
- Miscellaneous exercise ten

Situation One

An emergency services team is called to an area that has experienced strong winds, torrential rains and some flooding. In one place a bridge has been washed away and needs to be replaced to maintain an essential supply route. The team contacts an army engineering unit for assistance. The unit can bring up and lay a ready made pontoon bridge provided they know the width of the river the bridge has to span.

The emergency services team uses a direction compass and tape measure to measure the angles and distance shown in the diagram (i.e. ∠ABC = 90°, ∠CAB = 55° and AB = 20 metres).

Determine the width of the river.

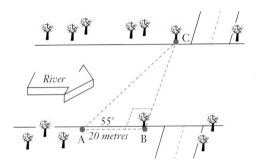

Situation Two

A company manufacturing steel frameworks is asked to quote a price for the manufacture and delivery of fifty roof frames like the one shown sketched below.

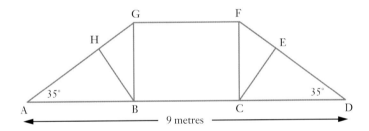

In order to quote a price for the job the company needs to know, amongst other things, the total length of steel required to make each frame. Determine the total length of steel required for each frame, add 10% for joints and wastage and then round up to the next whole metre.

Note: AD is horizontal, BG and CF are vertical, AB = BC = CD and ∠AHB = ∠DEC = 90°.

How did you get on with the situations on the previous page?

Did you think of drawing scale diagrams to determine the required lengths?

Perhaps instead you have encountered some *trigonometry* work in earlier years and remembered how to apply that to determine lengths of sides in right triangles.

Perhaps you used your calculator to determine lengths of unknown sides in right triangles.

In this chapter we will consider the use of *trigonometry* to determine sides and angles in right triangles.

Right angled triangles

The four triangles OAH, OBG, OCF and ODE shown below all have angles of 25°, 65° and 90°. As you know from unit one of this course, the four triangles are **similar**.

Each triangle is an enlargement, or reduction, of the others.

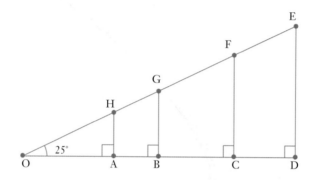

In the above diagram measure the length of ED and the length of OD and use your calculator to determine: $\dfrac{\text{length of ED}}{\text{length of OD}}$.

Did you get an answer of approximately 0.47?

On a piece of A4 paper accurately draw a large triangle with angles of 25°, 65° and 90°.

Label your triangle XYZ as shown in the diagram below.

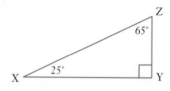

Measure the lengths of XY and YZ and determine: $\dfrac{\text{length of YZ}}{\text{length of XY}}$

Did you again find that your answer was approximately 0.47?

Ask others in your class what value they got for $\dfrac{\text{length of YZ}}{\text{length of XY}}$.

Is everyone getting an answer of approximately 0.47?

Even though each person's triangle may be a little smaller or larger than another person's all triangles with angles of 25°, 65° and 90° are similar to each other and are like photographic enlargements or reductions of each other.

Again as we know from unit one of this course, if two sides in a triangle are in a certain ratio then the corresponding sides in any similar triangle will also be in the same ratio.

Thus for the diagram on the previous page, the ratio of any two of the sides in $\triangle$OAH, e.g. $\dfrac{\text{HA}}{\text{OA}}$

will be the same as the ratio of the two corresponding sides in any of the other triangles,

$$\text{e.g.} \ \frac{\text{HA}}{\text{OA}} = \frac{\text{GB}}{\text{OB}} = \frac{\text{FC}}{\text{OC}} = \frac{\text{ED}}{\text{OD}}$$

and will be equal to the corresponding ratio in any other triangle with angles of 25°, 65° and 90°.

$$\text{i.e.} \ \frac{\text{HA}}{\text{OA}} = \frac{\text{GB}}{\text{OB}} = \frac{\text{FC}}{\text{OC}} = \frac{\text{ED}}{\text{OD}} = \frac{\text{YZ}}{\text{XY}} \approx 0.47$$

Any triangle with angles of 25°, 65° and 90° will give this same answer when the length of the side 'opposite the 25° angle' is divided by the length of the side 'between the 25° angle and the right angle'.

We call this ratio the **tangent** of 25°, abbreviated to **tan** 25°.

A more accurate value for tan 25° can be found from a calculator:

```
tan 25
                  0.4663076582
```

Trigonometry

The **tangent** of an angle is one of three ratios commonly used in the branch of mathematics called **trigonometry**. The three ratios are

- the **tangent** ratio (or **tan**)
- the **sine** ratio (or **sin**) and
- the **cosine** ratio (or **cos**).

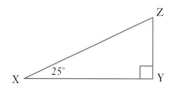

$$\tan 25° = \frac{ZY}{XY} \qquad \sin 25° = \frac{ZY}{XZ} \qquad \cos 25° = \frac{XY}{XZ}$$

The values of these ratios can be obtained from a scientific, or graphic, calculator.

Correct to two decimal places:

$\tan 25° = 0.47$

$\sin 25° = 0.42$

$\cos 25° = 0.91$

tan 25	
	0.4663076582
sin 25	
	0.4226182617
cos 25	
	0.906307787

Copy and complete the lines in the box below for △PQR shown.

Measure the lengths as accurately as possible and perform the division using your calculator. Then see how the answers you obtain compare with the values given above.

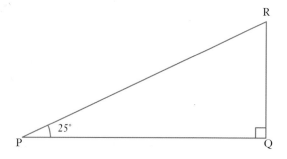

Side PQ is of length _____ cm

Side RQ is of length _____ cm

Side PR is of length _____ cm

$$\tan 25° = \frac{RQ}{PQ} = \text{\_\_\_\_}$$

$$\sin 25° = \frac{RQ}{PR} = \text{\_\_\_\_}$$

$$\cos 25° = \frac{PQ}{PR} = \text{\_\_\_\_}$$

Hypotenuse, opposite and adjacent

In a right triangle we call the side opposite the right angle the **hypotenuse**.

We then label the other two sides with respect to the angle we are considering:

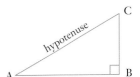

With respect to angle A we say that CB is the **opposite** side and AB is the **adjacent** side.

If, on the other hand, we are considering angle C it is now side AB that is the **opposite** side and BC is the **adjacent** side.

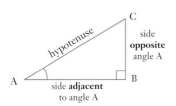

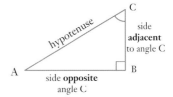

We then define the sine, cosine and tangent ratios as follows:

$$\sin x = \frac{\text{Opposite}}{\text{Hypotenuse}} \qquad \cos x = \frac{\text{Adjacent}}{\text{Hypotenuse}} \qquad \tan x = \frac{\text{Opposite}}{\text{Adjacent}}$$

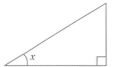

Hence

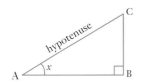

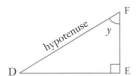

$$\sin x = \frac{\text{Opposite}}{\text{Hypotenuse}} = \frac{\text{CB}}{\text{AC}}$$

$$\cos x = \frac{\text{Adjacent}}{\text{Hypotenuse}} = \frac{\text{AB}}{\text{AC}}$$

$$\tan x = \frac{\text{Opposite}}{\text{Adjacent}} = \frac{\text{CB}}{\text{AB}}$$

$$\sin y = \frac{\text{Opposite}}{\text{Hypotenuse}} = \frac{\text{DE}}{\text{DF}}$$

$$\cos y = \frac{\text{Adjacent}}{\text{Hypotenuse}} = \frac{\text{EF}}{\text{DF}}$$

$$\tan y = \frac{\text{Opposite}}{\text{Adjacent}} = \frac{\text{DE}}{\text{EF}}$$

The sine, cosine and tangent ratios can be remembered using the mnemonic[*] **SOHCAHTOA**.

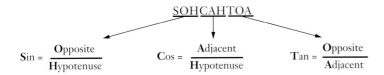

$$\text{Sin} = \frac{\textbf{O}\text{pposite}}{\textbf{H}\text{ypotenuse}} \qquad \text{Cos} = \frac{\textbf{A}\text{djacent}}{\textbf{H}\text{ypotenuse}} \qquad \text{Tan} = \frac{\textbf{O}\text{pposite}}{\textbf{A}\text{djacent}}$$

> **Note**
>
> [*]A mnemonic is a sequence of letters or words used to help us remember something.

The following examples show how these ratios can be used to determine unknown sides and angles in right angled triangles.

EXAMPLE 1

Find the value of x in each of the following, correct to one decimal place.

a

b

Solution

a We require the side **opposite** the 40° angle and we know the **adjacent**. Thus we use the tangent ratio because it involves these two sides.

$$\tan 40° = \frac{\text{Opp}}{\text{Adj}}$$

$$\therefore \quad \tan 40° = \frac{x}{5}$$

Multiply both sides by 5 to eliminate fractions.

$$5 \times \tan 40° = x$$

we write this as

$$5 \tan 40° = x$$

Thus $\quad x = 4.2$
(correct to 1 decimal place)

b We require the side **adjacent** to the 35° angle and we know the **hypotenuse**. Thus we use the cosine ratio because it involves these two sides.

$$\cos 35° = \frac{\text{Adj}}{\text{Hyp}}$$

$$\therefore \quad \cos 35° = \frac{x}{12}$$

Multiply both sides by 12 to eliminate fractions.

$$12 \times \cos 35° = x$$

we write this as

$$12 \cos 35° = x$$

Thus $\quad x = 9.8$
(correct to 1 decimal place)

or, using the solve facility on a calculator:

solve $\left(\tan(40) = \frac{x}{5}, x \right)$

$\{x=4.195498156\}$

solve $\left(\cos(35) = \frac{x}{12}, x \right)$

$\{x=9.829824531\}$

EXAMPLE 2

Find the value of x in each of the following, correct to one decimal place.

a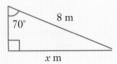

b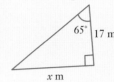

Solution

a We require the side **opposite** the 70° angle and we know the **hypotenuse**. Thus we use the sine ratio.

$$\sin 70° = \frac{\text{Opp}}{\text{Hyp}}$$

$$\therefore \qquad \sin 70° = \frac{x}{8}$$

i.e. $\quad 8\sin 70° = x$

Thus $\qquad x = 7.5$ (1 decimal place)

b We require the side **opposite** the 65° angle and we know the **adjacent**. Thus we use the tangent ratio.

$$\tan 65° = \frac{\text{Opp}}{\text{Adj}}$$

$$\therefore \tan 65° = \frac{x}{17}$$

i.e. $\quad 17\tan 65° = x$

Thus $\qquad x = 36.5$ (1 decimal place)

EXAMPLE 3

Find the value of x in each of the following, correct to one decimal place.

a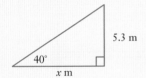

b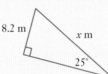

Solution

a We require the side **adjacent** to the 40° angle and we know the **opposite**. Thus we use the tangent ratio.

$$\tan 40° = \frac{\text{Opp}}{\text{Adj}}$$

$$\therefore \qquad \tan 40° = \frac{5.3}{x}$$

Multiply both sides by x

$$x\tan 40° = 5.3$$

Divide both sides by $\tan 40°$.

$$x = \frac{5.3}{\tan 40°}$$

Thus $\qquad x = 6.3$ (1 decimal place)

(Or alternatively use the solve facility

on a calculator to solve $\tan 40° = \frac{5.3}{x}$)

b We require the **hypotenuse** and we know the side **opposite** the 25° angle. Thus we use the sine ratio.

$$\sin 25° = \frac{\text{Opp}}{\text{Hyp}}$$

$$\therefore \qquad \sin 25° = \frac{8.2}{x}$$

Multiply both sides by x

$$x\sin 25° = 8.2$$

Divide both sides by $\sin 25°$.

$$x = \frac{8.2}{\sin 25°}$$

Thus $\qquad x = 19.4$ (1 decimal place)

(Or alternatively use the solve facility

on a calculator to solve $\sin 25° = \frac{8.2}{x}$)

EXAMPLE 4

Find the value of x in each of the following, to the nearest integer.

a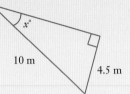

b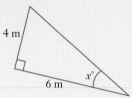

Solution

a We know the side **opposite** the required angle and we know the **hypotenuse**. Thus we use the sine ratio.

$$\sin x° = \frac{4.5}{10}$$
$$= 0.45$$

We require the angle whose sine is equal to 0·45.

Using 'inverse sine' or 'arc sine' on a calculator, often shown as $\sin^{-1}$, arcsin or perhaps ASIN we obtain

$$x = 27 \text{ (nearest integer)}$$

sin^{-1}(4.5÷10)
                26.74368395

b We know the side **opposite** the required angle and the side **adjacent** to the required angle. Thus we use the tangent ratio.

$$\tan x° = \frac{4}{6}$$
$$= 0.\overline{6}$$

We require the angle whose tangent is equal to $0.\overline{6}$.

Using 'inverse tan' or 'arc tan' on a calculator, often shown as $\tan^{-1}$, arctan or perhaps ATAN we obtain

$$x = 34 \text{ (nearest integer)}$$

tan^{-1}(4÷6)
                33.69006753

Notes regarding calculator usage

1   We can use the solve facility on some graphic calculators to solve equations such as $\sin x° = \frac{4.5}{10}$ and $\tan x° = \frac{4}{6}$. However, this needs care. In this chapter we are using sine, cosine and tangent in situations involving right triangles. We therefore know that when solving an equation like $\sin x = 0.45$ our answer must be between 0° and 90°. However, in more advanced mathematics (some of which we will see in the next chapter), meaning can be given to the sine, cosine and tangent of angles that are bigger than 90° and even to the sine, cosine and tangent of negative angles! Whilst there is only one value of x between 0° and 90° for which $\sin x = 0.45$ there are many angles outside this interval which have a sine equal to 0.45. The solve facility on a graphic calculator may give us one of these other answers instead. (See displays at the top of the next page.)

```
Eq: sin X = 4.5÷10
    X = 153.256316
  Lft = 0.45
  Rgt = 0.45
```

```
Eq: sin X = 4.5÷10
    X = 746.743684
  Lft = 0.45
  Rgt = 0.45
```

However, when an equation has more than one solution like this we can influence the one a calculator will give. On some calculators this is done by inputting an initial value of x and the calculator will tend to give the solution that is closest to this value. On other calculators we can instruct the calculator to only look for solutions in a particular range. In the display below left for example only solutions to the equation $\sin x = 0.45$ in the interval $0°$ to $90°$ are asked for whereas below right solutions in the range $0°$ to $180°$ are requested.

```
solve (sin(x)=0.45, x) │ 0≤x≤90
                {x=26.74368395}
```

```
solve (sin(x)=0.45, x) │ 0≤x≤180
            {x=153.256316, 26.74368395}
```

2   Some calculator programs allow the user to put in the known sides and angles of a triangle and, provided the information put in is sufficient, the program will determine the remaining sides and angles.

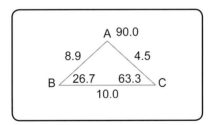

Other calculators can create a scale drawing of a geometrical figure and lengths and angles can then be determined from this drawing.

These programs can be useful but make sure that you understand the underlying ideas of sine, cosine and tangent and can reproduce the full method when required to do so.

Get to know your calculator.

Remember: When determining the lengths of sides in right triangles, the **Pythagorean theorem** can also be of use in situations where we are given the lengths of two sides and need to find the length of the third, as was mentioned in the *Preliminary work* section at the beginning of this book.

Applications

The previous examples all involved abstract triangles in which we had to determine an unknown side length or angle size. Some questions will be more applied and will refer to a particular situation in which a right angled triangle is involved. A simple, neat, clear diagram will then need to be drawn.

EXAMPLE 5

A ladder of length 5 metres leans against a vertical wall and just reaches the top of the wall. If the wall is 4.4 metres high calculate the angle the ladder makes with the horizontal ground (to the nearest degree) and the distance from the foot of the ladder to the wall (in metres correct to one decimal place).

Solution

First draw a diagram:

Knowing the hypotenuse and the side opposite the required angle we choose the sine ratio.

$$\sin x° = \frac{4.4}{5}$$

Solving gives $x \approx 62$

Using Pythagoras' theorem.

$$5^2 = 4.4^2 + y^2$$

Solving gives $y \approx 2.375$

The ladder makes an angle of 62° with the horizontal ground and the foot of the ladder is 2.4 metres from the wall.

Notice

- The final answers are not given as $x = 62$ and $y = 2.4$. The letters x and y were not part of the original question, we introduced them to help us obtain a solution. The final answer is given as a sentence that gives what was asked for.

- Having determined x, we could alternatively have then used 'tan' or 'cos' to determine the value of y:

$$\tan x° = \frac{4.4}{y}$$

Multiply both sides by y

$$y \tan x° = 4.4$$

$$\cos x° = \frac{y}{5}$$

Multiply both sides by 5

$$5 \cos x° = y$$

Being sure to use the accurate value of x, not the rounded value of 62, solving gives:

$y \approx 2·375$, as before. $y \approx 2·375$, as before.

Exercise 10A

1 Use your calculator to determine the following correct to 2 decimal places.

a $\sin 20°$ **b** $\cos 10°$ **c** $\tan 20°$ **d** $\tan 40°$

e $\tan 72°$ **f** $\cos 53.4°$ **g** $\sin 50°$ **h** $\cos 40°$

2 On a sheet of A4 paper accurately draw a large right triangle with angles of 35°, 55° and 90°. Measure the lengths of the three sides and using these measurements, and your calculator, determine estimates for

$$\sin 35°, \quad \cos 35°, \quad \tan 35°, \quad \sin 55°, \quad \cos 55°, \quad \tan 55°$$

and then check that your estimates are close to the accurate values the sin, cos and tan buttons on your calculator gives for these things.

3 Given that in each of the following x is one angle in a right triangle determine x in each case, giving your answer correct to one decimal place.

a $\sin x° = 0.2$ **b** $\cos x° = 0.4$ **c** $\tan x° = 1.3$ **d** $\sin x° = 0.3$

e $\cos x° = 0.25$ **f** $\sin x° = 0.8$ **g** $\tan x° = 2$ **h** $\cos x° = 0.9$

4 Determine the value of x in each of the following, giving your answers correct to one decimal place.

a $\sin 25° = \dfrac{x}{3}$ **b** $\cos 70° = \dfrac{x}{10}$ **c** $\tan 30° = \dfrac{x}{5}$

d $\sin 20° = \dfrac{7}{x}$ **e** $\cos 50° = \dfrac{9}{x}$ **f** $\tan 30° = \dfrac{7.3}{x}$

5 Given that in each of the following x is one angle in a right triangle determine x in each case, giving your answer correct to one decimal place.

a $\sin x° = \dfrac{2}{5}$ **b** $\cos x° = \dfrac{5}{7}$ **c** $\tan x° = \dfrac{7}{5}$

6 The right triangle shown on the right is scalene (i.e. the three sides of the triangle are of different lengths.)

Write each of the following in terms of two of a, b and c.

a $\sin P$ **b** $\cos P$ **c** $\tan P$

d $\cos Q$ **e** $\sin Q$ **f** $\tan Q$

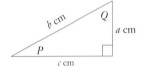

7 Which of the following statements is true for the right triangle shown on the right?

$$q^2 = p^2 + r^2 \qquad p^2 = q^2 + r^2 \qquad r^2 = p^2 + q^2$$

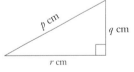

8 In each of the following determine x by:

i   accurately drawing the triangle   **ii**   using trigonometry.

a

x cm, 37°, 7 cm

b

3 cm, $x°$, 5 cm

Find the value of *x* (and *y* if applicable) in each of numbers **9** to **30** clearly showing your use of trigonometry or Pythagoras in each one. (Give answers correct to one decimal place if rounding is necessary.)

9

10
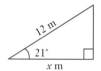

11
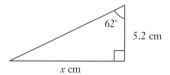

12
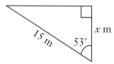

13
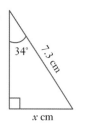

14

15
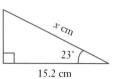

16
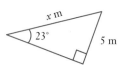

17
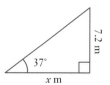

18
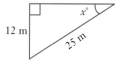

19
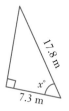

20
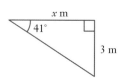

21

22

23

x m, 71°, 2.1 m

24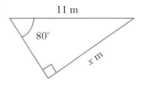

11 m, 80°, x m

25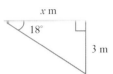

x m, 18°, 3 m

26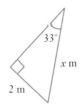

33°, x m, 2 m

27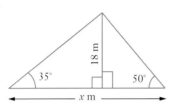

35°, 18 m, 50°, x m

28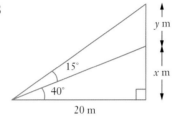

y m, 15°, x m, 40°, 20 m

29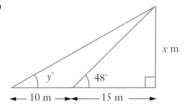

x m, y°, 48°, 10 m, 15 m

30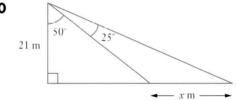

50°, 25°, 21 m, x m

31 Triangle ABC is right angled at B. If AC = 17.6 cm and ∠CAB = 32° find
  a the length of AB, in centimetres correct to one decimal place,
  b the length of BC, to the nearest millimetre.

32 Triangle DEF is right angled at D. If ED = 7 cm and FD = 5 cm find
  a the size of ∠FED, to the nearest degree,
  b the length of FE, to the nearest millimetre.

33 The diagram shows a ladder leaning against a vertical wall and making an angle of 62° with the horizontal ground.

If the ladder is 8 metres in length calculate
  a how high the ladder reaches up the wall, to the nearest centimetre,
  b the horizontal distance from the foot of the ladder to the wall, to the nearest centimetre.

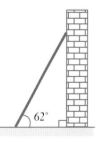

62°

34 The outdoor light that illuminates the driveway of a two-storey house has a light globe that needs replacing. A ladder of length 5 metres is placed with its foot on the horizontal ground and 2 metres from the vertical wall of the house. In this position the ladder just reaches the light.

Find

a the angle the ladder makes with the ground, to the nearest degree,

b the height of the light above the ground, in metres and correct to one decimal place.

35 A person flying a kite holds the line 1 metre above level ground and has 45 metres of line out. If the line is straight and makes 62° with the horizontal what is the height of the kite above ground level (to the nearest metre)?

36 To reduce the force acting on the end of a garden fence due to the wind the fence can be 'raked down'.

The diagram on the right shows a fence raked down from a height of 1.8 metres to 1 metre in a horizontal distance of 2 m.

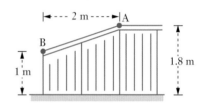

Find

a the acute angle AB makes with the horizontal, to the nearest degree.

b the length of AB, to the nearest centimetre.

37 The diagram on the right shows a simple bridge design.

AD and BC are horizontal. FB and EC are vertical.

∠BAF = ∠CDE = 50°. AD = 24 m and AF = FE = ED.

Calculate the lengths of AB, BF and FC giving all answers to the nearest centimetre.

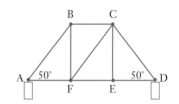

38 The diagram shows a road bridge that can be opened to allow tall ships to pass underneath.

AB = AE = 20 metres and *h* is the distance from C, the mid-point of AB, to D, the point on the bridge vertically above C.

If *h* needs to be 8 m find θ in degrees correct to one decimal place. For this value of θ find the length of AD as a percentage of AE (to the nearest percent).

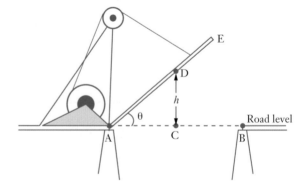

ISBN 9780170390262

39 A vertical mast stands on level ground and is supported by a number of
wires, as shown in the diagram. All these wires have one end attached to
the ground, six metres from the base of the mast, and their other ends are
attached to points that are either one-third or two-thirds of the way up
the mast.

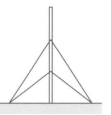

If the height of the mast is fifteen metres, find:

a the length of one of the 'short wires' (nearest centimetre) and the angle it makes with the
ground (nearest degree),

b the length of one of the 'long wires' (nearest centimetre) and the angle it makes with the
ground (nearest degree).

40 The diagram on the right shows the timbers forming
part of a roof. The framework is symmetrical, AE is
horizontal, HB, GC and FD are vertical and
∠BAH = 40°. AH = HG = GF = FE = 2 m.

Find the length of AC, CG, BH and HC giving your
answers to the nearest centimetre.

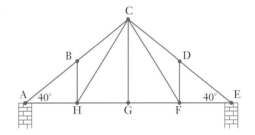

41 A vertical pole of height 20 metres stands on horizontal ground and is supported by a number of guy
wires. Each wire has one end attached to a point three-quarters of the way up the pole and the other
end attached to one of the fastenings situated on the ground, 8 m from the base of the pole. Find the
acute angle each wire makes with the horizontal, giving your answer to the nearest degree.

42 A mast AD is to stand vertically on horizontal ground. Part of the mast,
CD in the diagram, is to be below ground. A straight support wire has one
end fastened to the ground, at point E in the diagram, and the other to a
point B on the mast. Angle BEC is to be no less than 42° and no more than
52° and BC will be no less than 9·9 metres and no more than 10.2 metres.

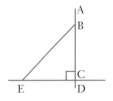

Based on these figures determine

a the largest possible length of the support wire EB (nearest cm),

b the shortest possible length of the support wire EB (nearest cm).

43 A pendulum of length 80 cm swings 15° either side of the vertical.

What is the vertical rise the bob of the pendulum makes above its lowest position, to the nearest millimetre?

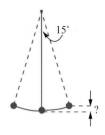

44 The diagram shows a mobile crane lifting an 8 metre pole into the vertical position. The cable from the crane is attached to a point C where AC is four fifths of AB. At the instant shown in the diagram CD = 8 metres, how high is point D above the horizontal ground (to the nearest metre)?

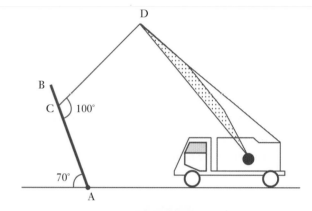

45 (Challenge.)

The framework shown below is to be made out of lengths of steel.

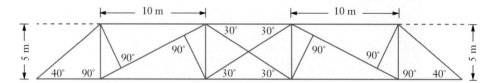

The framework consists of a right angled triangle on each end with three rectangles in the middle.

The company contracted to make it needs to know the total length of steel required.

Find the length of steel required, to the nearest whole metre.

46 (Challenge? Maybe, but hint makes it okay.)

Given the diagram shown on the right determine the values of x and y giving each answer correct to one decimal place.

Hint: Obtain the height in terms of y from one triangle, and then find the height in terms of y from another triangle and then …

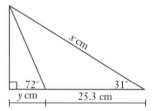

ISBN 9780170390262

Accuracy and trigonometry questions

So far all of the examples and almost all of the questions in this chapter have stated the degree of rounding that the answers should be given to. If this is not stated you should 'round appropriately'. Just what is appropriate depends upon the accuracy of the data we are given and what is appropriate for the situation. For example, if a question gives us a right triangle with one side of length 3.2 cm and one angle of size 36°, our calculator may give us the length of some other side as being 2.32493609 but it would be quite inappropriate to claim this sort of accuracy because it is far beyond the accuracy of the information used to obtain it.

In general, if we are not told the accuracy to give an answer to, our final answer should not be more accurate than the accuracy of the data we use to obtain it. If a question gives us a length in cm, to 1 decimal place, we should not claim greater accuracy for any lengths we determine. Sometimes we may need to use our judgement of the likely accuracy of the given data. Given a length of 5 cm we might assume this has been measured to the nearest mm and hence give answers similarly to the nearest mm. (Theoretically a measurement of 5 cm measured to the nearest mm should be recorded as 5.0 cm but this is often not done.)

In situations where accuracy is crucial any given measurements could be given with 'margins of error' included, for example 3.2 cm ± 0.05 cm, 36° ± 0.5°. More detailed error analysis could then be carried out and the margins of error for the answer calculated. However this is beyond the scope of this text and, as mentioned earlier: *If a question does not state the degree of rounding required your final answers should be rounded 'appropriately'.*

You are already used to rounding appropriately in some situations not involving trigonometry. For example, if asked for the sale price in an '8% off everything sale' for something usually costing $25.45 you would give the answer as $23.40, not the $23.414 value a calculator gives for $25·45 × 0.92. If asked how many chocolate bars costing $1.40 each we could purchase with $8 we would not give the calculator answer of 5.714285714 bars, even though we would know that the values of $1.40 and $8 were exact. Instead we would say that 5 bars could be purchased and, if we wanted to give more information, we could add that $1 change would be given.

Some questions requiring the use of trigonometry involve situations that mention **bearings** and/or **angles of elevation** or **depression**. It is often these concepts that cause errors more so than the trigonometry itself. The next few pages cover these concepts.

Bearings

From one location, the direction we would need to travel to reach a second location can be given as a bearing. These are angles, expressed as three figures, and are measured from North, clockwise, as shown below.

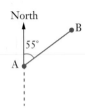

Bearing of B from A is 055°

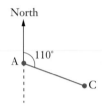

Bearing of C from A is 110°

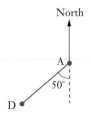

Bearing of D from A is 230°

Alternatively, bearings can be expressed as a certain number of degrees (between 0 and 90) to the East or West of North or South. Thus the above *three figure bearings* could be expressed as the following *compass bearings*:

Bearing of B from A is	Bearing of C from A is	Bearing of D from A is
N 55° E	S 70° E	S 50° W

EXAMPLE 6

From town A, town B lies 7.2 km away on a bearing of 070°.

From town B, town C lies 8.4 km away on a bearing of 160°.

Find the distance and bearing of C from A.

Solution

First make a sketch of the situation:

Note that with the given bearings ∠ABC = 90°.

In △ABC, by Pythagoras' theorem, $\qquad AC^2 = 7.2^2 + 8.4^2$

Thus $\qquad\qquad\qquad\qquad\qquad AC \approx 11.06$ km

Also $\qquad\qquad\qquad\qquad\qquad \tan \angle BAC = \dfrac{8.4}{7.2}$

giving $\qquad\qquad\qquad\qquad\qquad \angle BAC \approx 49.4°$

Thus the bearing of C from A is approximately 119°, i.e. (49° + 70°).

Town C is 11.1 km from A, on a bearing of 119°.

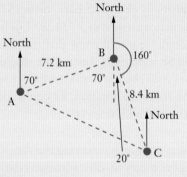

Elevation and depression

- **Angles of elevation** are measured from the horizontal, up.

- **Angles of depression** are measured from the horizontal, down.

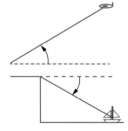

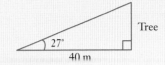

EXAMPLE 7

From a point on level ground, 40 metres from a tree, the angle of elevation of the top of the tree is 27°. Calculate the height of the tree.

Solution

First make a sketch of the situation.

With respect to the 27° we know the length of the adjacent side and require the length of the opposite side. Thus we choose the tangent ratio.

$$\tan 27° = \frac{\text{height of tree}}{40}$$

Solving gives: height of tree ≈ 20.4 metres

The height of the tree is approximately 20 metres.

More vocabulary

• Note also that if a question refers to a line **subtending** an angle at a point this is the angle formed by joining each end of the line to the point.

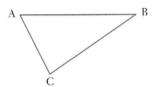

Line AB subtends
∠ACB at C.

Chord DE subtends
∠DFE at the centre, F.

• If points are referred to as being **collinear** this means they lie in a straight line.

Exercise 10B

1 From the diagram on the right find the bearing of:

a	B from A	**b**	C from A
c	D from A	**d**	E from A
e	F from A	**f**	G from A
g	A from B	**h**	A from C
i	A from D	**j**	A from E
k	A from F	**l**	A from G.

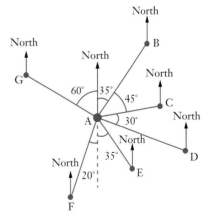

2 a What is the angle of elevation of the aeroplane from A?

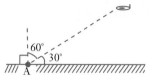

b What is the angle of depression of the ship from B?

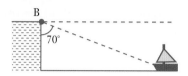

c What is the angle of depression of point C from the top of the tower?

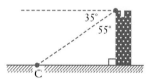

d What is the angle of elevation of the top of the flagpole from D?

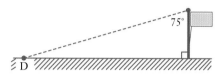

3 When the angle of elevation of the sun is 28° a vertical flag pole casts a shadow of length 22.4 metres on horizontal ground. Calculate the height of the flagpole.

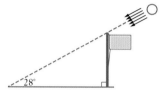

4 Find the angle of elevation of the sun if a 2.0 metre pole held vertically on horizontal ground casts a shadow of length 4.1 metres. Give your answer correct to the nearest degree.

5 A flagpole stands vertically on level ground. When the sun's elevation is 24° the flagpole casts a shadow of length 22.5 metres. Find the height of the flagpole.

6 A and B are two points on horizontal ground. A mast of length 540 cm is to stand vertically with its base at B. From A, the top of the mast will have an angle of elevation of 17°. A straight wire is to run from the top of the mast to the point A. How far is this, rounded up to the next metre?

7 At 9 a.m. one morning two ships leave a harbour and head out to sea. One ship travels at a steady 4 km/h on a bearing 110° and the other ship maintains 5 km/h on a bearing 200°. To the nearest kilometre how far apart are the ships one and a half hours later?

Shutterstock.com/studio trebuchet

8 The three points A, B and C lie on horizontal ground and form a straight line with B between A and C. A vertical tower of height 40 metres stands at C. The angle of elevation of the top of the tower is 18° from A and 35° from B. How far is B from A (to the nearest metre)?

ISBN 9780170390262

9 From ship A, ship B lies 12.2 km away on a bearing N 58° W. From ship B, ship C lies on a bearing S 32° W. If the bearing of C from A is S 59° W how far is ship C from ship A?

10 Three collinear points A, B and C lie on horizontal ground with B between A and C. A vertical tower of height 42 metres stands at B. The angle of elevation of the top of the tower is 28° from A and 17° from C. How far is C from A (to the nearest metre)?

11 Three collinear points A, B and C lie on horizontal ground with B between A and C. A vertical tower of height 36 metres stands at C. The angle of elevation of the top of the tower is 15° from A and 40° from B. How far is B from A (to the nearest metre)?

12 Two vertical towers stand on level ground. From the top of one tower, of height 40 metres, the top and base of the second tower have angles of elevation and depression of 20° and 30° respectively. Find the height of the second tower.

13 A tree stands vertically on a hillside that is inclined at 20° to the horizontal. When the angle of elevation of the sun is 39° (i.e. 39° with the horizontal) the tree casts a shadow of length 35.3 metres straight down the slope. How tall is the tree?

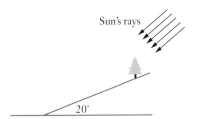

14 A forest warden on fire look-out duty in an observation tower notices smoke directly North of his position. From a second tower, situated 5.3 km due East of the first, another warden sees the smoke on a bearing 335°. How far is the smoke from the first observation tower?

15 An observer in an aircraft flying at an altitude of 500 metres notices two ships at sea. At the moment the observer sees the ships as being 'in line' he records their angles of depression as 30° and 40° respectively. How far apart are the ships?

16 A vertical flagpole stands on top of a vertical tower of height 40 m. At a point level with the base of the tower and 60 m from it, the flagpole subtends an angle of 10°. How long is the flagpole?

17 A and B are two points on level ground, 19.6 metres apart. A vertical flagpole at B subtends an angle of 40° at the eye of a person standing at A and whose 'eye height' is 1.6 m. Find the height of the flagpole.

18 From a point on level ground the angle of elevation of a vertical flagpole is 40°. From the same position find the angle of elevation of the point three-quarters of the way up the flagpole.

Miscellaneous exercise ten

This miscellaneous exercise may include questions involving the work of this chapter, the work of any previous chapters, and the ideas mentioned in the Preliminary work section at the beginning of the book.

1 Determine the value of x in each of the following, giving your answers correct to one decimal place.

a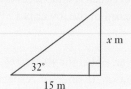

b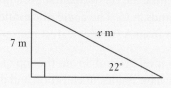

c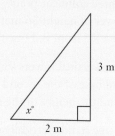

d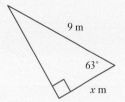

e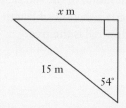

f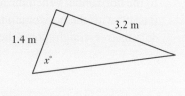

2 Find the mean and the median of the following six amounts:

$13 600 $5700 $23 400 $2100 $14 600 $98 700

3 The 35 students in a class sat a test that was marked out of 40. The 20 boys had a mean score of 24.35 and the class mean was 23. What was the mean score of the girls in the class?

4 Solve the following equations.

    a $7x - 15 = 132$ **b** $3(2x - 1) + 2x = 17$

    c $\dfrac{2x + 1}{3} = 5$ **d** $\dfrac{5}{x} = 8$

5 Formula: $v^2 = u^2 + 2as$

    a   Find s given that $v = 13$, $u = 5$ and $a = 24$.

    b   Find a given that $v = 21$, $u = 17$ and $s = 19$.

6 I think of a number, multiply it by three, add seven and then divide the answer by two. At the end of all this the number I end up with is eleven more than the number I first thought of. Find the number first thought of.

7 John takes out a loan which involves simple interest charged at the rate of 7.5% per annum. After 4 years John repays $11 180 which clears the loan and interest.

How much did John borrow in the first place?

8 The 'rectapatterns' shown below all follow the pattern shown on the right.

Copy and complete the eight 'rectapatterns' shown below.

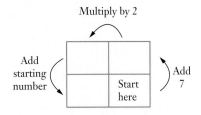

Multiply by 2

Add starting number

Add 7

Start here

a

5

b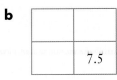

7.5

c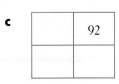

92

d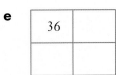

17

e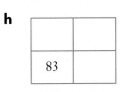

36

f

14

g

59

h

83

9 The diagram below shows the percent of total national income earned by each tenth of the population of a particular country in one year, richest at the top.

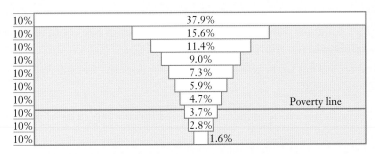

10%	37.9%
10%	15.6%
10%	11.4%
10%	9.0%
10%	7.3%
10%	5.9%
10%	4.7%
10%	3.7%
10%	2.8%
10%	1.6%

Poverty line

Source of data: The New Internationalist Magazine.

For the particular year and country involved:

a what percentage of the population lie below the poverty line?

b what is the '37.9%' in the above graph telling you?

10 Find the equation of the straight line with gradient 0.5, passing through $(3, 4)$.

Each of the points $F(9, f)$, $G(-9, g)$, $H(h, 9)$, $I(i, 1.5)$ and $J(3.8, j)$ lie on this line.

Determine the values of f, g, h, i and j.

11 A group of students sat an exam. The mean score for the boys was 56% and for the girls was 62%.

a If the group had the same number of girls as it had boys what would be the mean of the whole group?

b If the mean for the whole group was actually 59.8% were there more boys than girls in the group or were there more girls than boys?

12 Find the equations of the lines A to J shown below.

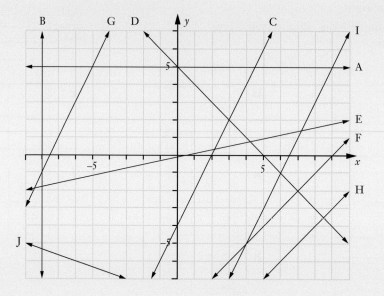

13 Scientists investigating levels of pollution in a particular stretch of river estimate that N, the number of fish that part of the river can support, depends on P, the number of tonnes of pollutant in that part of the river, according to the rule:

$$N \approx 60\,000 - 2100P$$

a How many fish does this rule suggest this stretch of the river can support if $P = 5$.

b If the level of pollution reaches 18 tonnes what number of fish could this stretch of the river support according to the above rule.

c If fish numbers are not to drop below 45 000 what does the formula suggest the pollution level must not exceed?

14 The diagram below shows a headlamp beam adjusted down to avoid dazzling oncoming drivers.

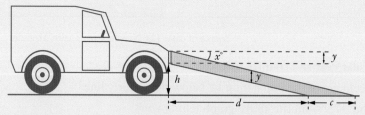

The light beam is angled at $x°$ below the horizontal and the distances h, d, c and y are as shown in the diagram. Assuming that the beam does not spread out at all and that the ground is horizontal, find d and c given that:

$$x = 4, h = 80 \text{ cm and } y = 20 \text{ cm.}$$

ISBN 9780170390262

11.

Trigonometry for triangles that are not right angled

- Area of a triangle
- Triangles that are not right angled
- Area of a triangle given two sides and the included angle
- The sine rule
- The cosine rule
- Miscellaneous exercise eleven

Area of a triangle

From unit one of this course, and from earlier years, you are already familiar with the formula for the area of a triangle:

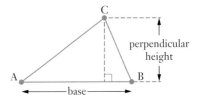

$$\text{Area of a } \triangle = \frac{1}{2} \times \text{base} \times \text{perpendicular height}$$

Suppose you are asked to determine the area of a triangle for which you do *not* know the perpendicular height but instead know two sides and the angle between them, as shown on the right. How would you proceed then?

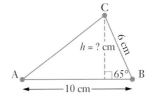

Exercise 11A

1 Find *h* in triangle ABC shown above and hence determine the area of the triangle.

Find the areas of each of the following triangles (not drawn to scale).

2
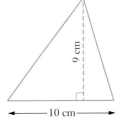

3
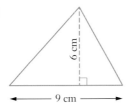

4
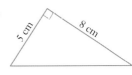

5
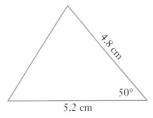

6
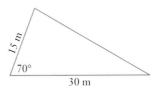

7
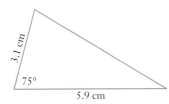

8
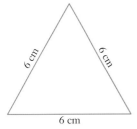

9
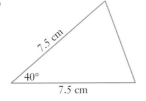

10
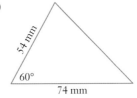

Triangles that are not right angled

On the previous page we were asked, amongst other things, to determine the area of △ABC, given the lengths of BA and BC and the size of ∠ABC. (i.e. given two sides and the angle between them.)

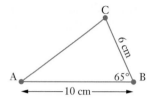

By drawing the perpendicular from C to AB we obtain right triangles. This allows trigonometry to be used to determine the height, and hence the area, of △ABC.

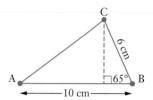

(Alternatively △ABC could be drawn accurately from the given information and the perpendicular height could be measured.)

This approach of drawing the perpendicular from one vertex to the opposite side allows trigonometry to be used for a triangle that is not right angled. We will use this approach in this chapter to obtain three formulae that are useful when dealing with triangles that are not right angled.

We will consider:
- a formula for the area of a triangle,
- the sine rule formula,
- the cosine rule formula.

Note: In obtaining these formulae we will use the usual convention for naming the sides and angles of a triangle. i.e. in triangle ABC, the three angles are labelled A, B and C according to their vertex and the sides opposite these angles are labelled a, b and c respectively.

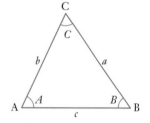

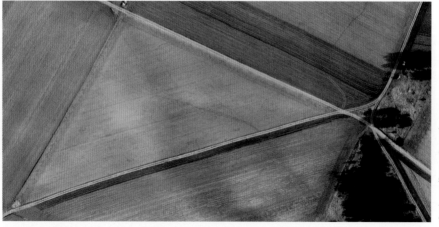

Area of a triangle given two sides and the included angle

Consider △ABC with the perpendicular from B to AC meeting AC at D (see diagram).

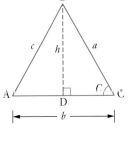

In △BDC, $$\sin C = \frac{h}{a}$$

Multiplying by a to isolate h, $$h = a \sin C$$

Using $$\text{Area of triangle} = \frac{1}{2} \text{ base} \times \text{height}$$

we have $$\text{Area of triangle} = \frac{1}{2} \times b \times a \sin C$$

Thus:

$$\text{Area of a triangle} = \frac{ab \sin C}{2}$$

i.e. the area of a triangle is half the product of two sides multiplied by the sine of the angle between them.

EXAMPLE 1

Find the area of the triangle shown.

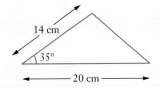

Solution
$$\text{Area} = \frac{20 \times 14 \times \sin 35°}{2}$$

$$\approx 80.3 \text{ cm}^2$$

Now that we are dealing with any triangle, not just right triangles, we could have an obtuse angle between the two sides of known length, as shown below.

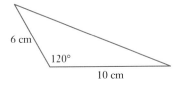

Applying our formula will involve the sine of an obtuse angle:

$$\text{Area} = \frac{10 \times 6 \times \sin 120°}{2}$$

How does your calculator respond when asked for sin 120° or sin 130° or sin 140° …?

Did you notice that

$$\sin 120° = \sin 60°$$
$$\sin 130° = \sin 50°$$
$$\sin 140° = \sin 40°$$

in fact, to generalise: $\sin(180° - C) = \sin C$?

sin 120	
	0.8660254038
sin 60	
	0.8660254038
sin 130	
	0.7660444431
sin 50	
	0.7660444431

This fact means that we can use our area formula,

$$A = \frac{1}{2} ab \sin C$$

for all triangles, even those for which angle C is obtuse.

Consider the acute angled $\triangle ABC$ below left and the obtuse angled $\triangle ABC$ below right.

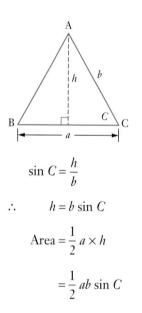

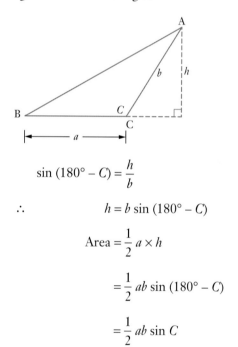

$$\sin C = \frac{h}{b}$$

$$\therefore \qquad h = b \sin C$$

$$\text{Area} = \frac{1}{2} a \times h$$

$$= \frac{1}{2} ab \sin C$$

$$\sin (180° - C) = \frac{h}{b}$$

$$\therefore \qquad h = b \sin (180° - C)$$

$$\text{Area} = \frac{1}{2} a \times h$$

$$= \frac{1}{2} ab \sin (180° - C)$$

$$= \frac{1}{2} ab \sin C$$

Thus for all triangles our area formula $A = \frac{1}{2} ab \sin C$ applies.

However, the fact that $\sin(180° - C) = \sin C$ does present a difficulty if a question gives the area of a triangle and the lengths of two sides of the triangle and asks for the size of the angle between the two sides of known length. Which answer do we give – the acute angle or the obtuse angle?

For example, suppose $\triangle ABC$ has an area of 3 cm^2 and is such that $a = 4$ cm and $b = 3$ cm.

Using   Area $= \frac{1}{2} ab \sin C$

$$3 = \frac{1}{2} \times 4 \times 3 \sin C$$

$$3 = 6 \sin C$$

and so $\sin C = 0.5$

sin 150	
	0.5
sin 30	
	0.5

Now comes our dilemma:

Does $C = 30°$ or does $C = 150°$?

The dilemma is genuine because for the information we are given there *are* two possible triangles that 'fit the facts':

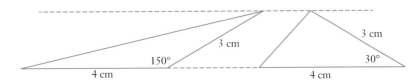

The two triangles each have a base of 4 cm and are the same height as each other. Hence their areas will indeed be equal, and in this case each equal to 3 cm².

The information we have been given about triangle ABC is said to be *ambiguous*. (The word ambiguous meaning open to more than one interpretation.) However, there is no need to panic. In this unit, we will be given sufficient information for such ambiguity to be avoided, as in the next example.

EXAMPLE 2

Triangle DEF is such that $e = 16$ cm, $d = 12$ cm, the area of the triangle is 78 cm² and $\angle DFE$ is an acute angle. Find the size of $\angle DFE$ giving your answer to the nearest 0.1°.

Solution

Using   Area $= \frac{1}{2} ab \sin C$

$$78 = \frac{1}{2} \times 12 \times 16 \sin \angle DFE$$

∴ $\sin \angle DFE = 0.8125$

and so, given that $\angle DFE$ is an acute angle, using a calculator:

$\angle DFE = 54.3°$ (to nearest 0.1°)

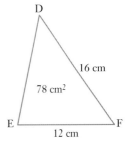

(If instead we had been told that $\angle DFE$ was an obtuse angle our final answer would have been 125.7°, i.e. 180° − 54.3°. Had we not been told anything about $\angle DFE$ then two possible triangles would exist that each satisfied the given facts.)

Given that the triangle sketched on the right has an area of 7 cm^2 find x correct to one decimal place.

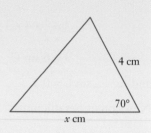

Solution

$$\text{Area} = \frac{1}{2}(x)\, 4 \sin 70°$$

$$\therefore\ 7 = \frac{1}{2}(x)\, 4 \sin 70°$$

i.e. $7 = 2\,x \sin 70°$

Solving this equation gives $x = 3.7$, correct to 1 decimal place

Note: It is also possible to determine the area of a triangle, given the lengths of the three sides of the triangle, using a result known as **Heron's 's' formula**:

$$\text{Area of } \triangle ABC = \sqrt{s(s-a)(s-b)(s-c)} \quad \text{where } s = \frac{a+b+c}{2}.$$

One of the questions of a later exercise in this chapter, and one of the questions in a later Miscellaneous Exercise, reminds you of this formula and requires you to use it.

Exercise 11B

Find the area of each triangle in questions 1 to 11, giving your answers in square centimetres and correct to one decimal place. (Diagrams not necessarily to scale).

1
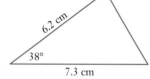

2
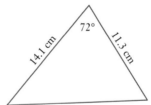

3
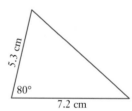

4
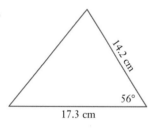

5
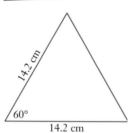

6
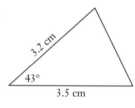

7
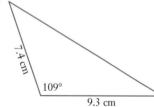

8
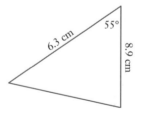

9
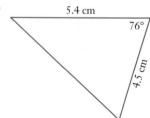

10 $\triangle ABC$ given that $AB = 8$ cm, $BC = 7$ cm, $AC = 5$ cm and $\angle BAC = 60°$.

11 $\triangle PQR$ given that $PQ = 7$ cm, $PR = 8$ cm, $RQ = 3$ cm and $\angle PRQ = 60°$.

Find the value of x in each of the following, correct to one decimal place, given that the area of each triangle is as stated. (Diagrams not necessarily drawn to scale.)

12

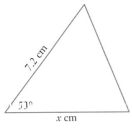

Area = 19.6 cm^2

13

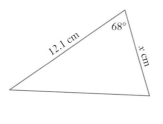

Area = 40.9 cm^2

14

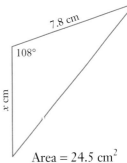

Area = 24.5 cm^2

Find the size of $\angle ABC$ in each of the following, correct to the nearest degree, given that each triangle is acute angled and the area of each triangle is as stated. (The diagrams are not necessarily drawn to scale.)

15

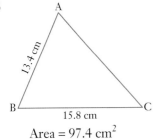

Area = 97.4 cm^2

16

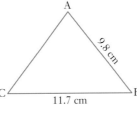

Area = 45.2 cm^2

17

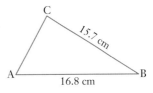

Area = 69.9 cm^2

18 If farming land in a particular region costs $12 300 per hectare find the cost of each of the following areas, to the nearest $1000. (1 hectare = 10 000 m^2.)

a

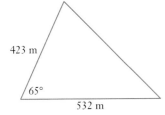

b

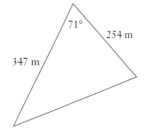

19 A triangular block of land has two sides of lengths 45 m and 30 m and the angle included between them is 70°. A second triangular block has two sides of lengths 48 m and 35 m and the angle included between them is 50°.

Which block has the greater area and by how much (to the nearest square metre)?

20 The owners of two neighbouring triangular blocks of land, shown as A and B in the diagram on the right, are offered a total of $1 250 000 by a property developer for the two blocks together. If they were to accept this offer and divide the money between them in the ratio of the land areas of the blocks how much would each owner receive?

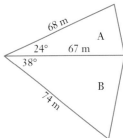

21 (Note: A square of side 100 metres has an area of 1 hectare.)

A farmer wishes to lease one hectare of his land to an investor who wishes to use it to grow Tasmanian Blue Gum trees. The investor intends harvesting these fast growing trees and selling the wood to a paper making company as woodchip. The farmer, for his part, simply has to fence off suitable land for the investor to use.

Rather than having to use new fencing around the whole area the farmer chooses a triangular site that allows existing fencing to be used on two sides (AB and AC in the diagram). The farmer measures the distance AB as 173 metres and measures ∠CAB as 40°. He wishes to locate point C so that △ABC will have an area of one hectare. He asks you to calculate the length AC for him. Calculate this length, rounding your answer *up* to the next whole metre.

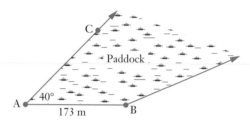

The sine rule

If we were given the triangle on the right and were asked to find *a* we could accurately draw the triangle and measure the required length. However, great accuracy is not easy to achieve and drawing can be time consuming. Once again we could proceed by drawing the perpendicular from B to AC so that our right triangle trigonometry work can be used, as shown below.

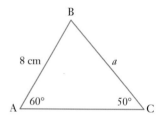

Draw the perpendicular from B to meet AC at D (see diagram).

In △ABD, $\sin 60° = \dfrac{h}{8}$

            ∴ $h = 8 \sin 60°$

In △BCD, $\sin 50° = \dfrac{h}{a}$

            ∴ $\sin 50° = \dfrac{8 \sin 60°}{a}$

Solving gives $a \approx 9.04$ cm

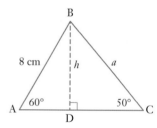

$$\text{solve}\left(\sin(50) = \frac{8 \cdot \sin(60)}{x}, x\right)$$

$$\{x = 9.044126999\}$$

As we will see on the next page, if we apply this technique to a general triangle ABC we obtain the **sine rule**:

$$\frac{a}{\sin A} = \frac{b}{\sin B} = \frac{c}{\sin C}$$

Consider a triangle ABC as shown below left for an acute angled triangle and below right for an obtuse angled triangle.

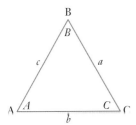 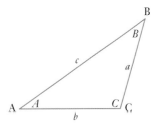

Drawing the perpendicular from B to meet AC at D:

Drawing the perpendicular from B to meet AC produced at D:

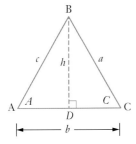

From $\triangle$ABD: $\sin A = \dfrac{h}{c}$

$\therefore$ $h = c \sin A$ [1]

From $\triangle$CBD: $\sin C = \dfrac{h}{a}$

$\therefore$ $h = a \sin C$ [2]

From $\triangle$ABD: $\sin A = \dfrac{h}{c}$

$\therefore$ $h = c \sin A$ [1]

From $\triangle$CBD: $\sin (180° - C) = \dfrac{h}{a}$

$\therefore$ $h = a \sin C$ [2]

Thus for both the acute triangle and the obtuse triangle:

From [1] and [2] $c \sin A = a \sin C$

Thus $\dfrac{c}{\sin C} = \dfrac{a}{\sin A}$ [3]

If instead we draw the perpendicular from A to BC we obtain

$$\dfrac{b}{\sin B} = \dfrac{c}{\sin C}$$ [4]

From [3] and [4] it follows that

$$\dfrac{a}{\sin A} = \dfrac{b}{\sin B} = \dfrac{c}{\sin C}$$

This is the **sine rule**.

Hint

Rather than learning this formula, notice the pattern:

Any side on the sine of the opposite angle is equal to any other side on the sine of its opposite angle.

The sine rule–finding lengths of sides

EXAMPLE 4

Find the value of x in the following, giving answers correct to one decimal place.

a

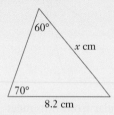

60°

x cm

70°

8.2 cm

b

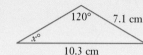

120°   7.1 cm

$x°$

10.3 cm

Solution

a   By the sine rule

$$\frac{x}{\sin 70°} = \frac{8.2}{\sin 60°}$$

Multiply by sin 70° to isolate x

$$x = \frac{8.2 \sin 70°}{\sin 60°}$$

$$= 8.9 \text{ (to 1 decimal place)}$$

b   By the sine rule

$$\frac{10.3}{\sin 120°} = \frac{7.1}{\sin x°}$$

Multiply by $(\sin x°)(\sin 120°)$

$$10.3 \sin x° = 7.1 \sin 120°$$

$$\therefore \quad \sin x° \approx 0.5970$$

$$x = 36.7 \text{ (to 1 decimal place)}$$

Or, using the 'solve' ability of some calculators:

$$\text{solve}\left(\frac{x}{\sin(70)} = \frac{8.2}{\sin(60)}, x\right)$$

$$\{x = 8.897521316\}$$

$$\text{solve}\left(\frac{10.3}{\sin(120)} = \frac{7.1}{\sin(x)}, x\right) \Big| 0 \le x \le 180$$

$$\{x = 143.346877, x = 36.65312298\}$$

Note:

• In part **b** we went from sin $x° \approx 0.5970$ to $x = 36.7$ (to 1 decimal place) despite there being another value of x between 0 and 180 for which sin $x° = 0.5970$, and that is $(180 - 36.7)$, i.e. 143.3, as the calculator shows when asked for solutions in the interval $0 \le x \le 180$. However, in the given triangle x cannot be 143.3 because the triangle already has one obtuse angle and cannot have another. However, whilst this may not always be the case, and an *ambiguous* situation could occur when both answers are possible, this unit will not include such situations and sufficient information will be given to be able to dismiss one of the solutions.

ISBN 9780170390262

As was mentioned in the previous chapter on right triangles, some calculator programs allow the user to put in the known sides and angles of a triangle and, provided the information put in is sufficient, the program will determine the remaining sides and angles.

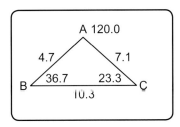

Also some calculators allow us to create a scale drawing of the triangle and find lengths and angles that way.

These programs can be useful but make sure that you understand the underlying idea of the sine rule (and the cosine rule which we will see later in this chapter) and can demonstrate the appropriate use of these rules when required to do so.

EXAMPLE 5

The sine rule–finding angles

Find the value of x in the triangle shown on the right.

(Give the answer correct to one decimal place.)

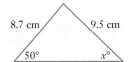

Note

Note first that $x°$, being opposite a side of length 8.7 cm, must be less than the 50° which is opposite a side of length 9.5 cm. (For any two sides of a triangle, the larger of the two sides has the larger opposite angle.)

Solution

By the sine rule:

$$\frac{9.5}{\sin 50°} = \frac{8.7}{\sin x°}$$

Multiply by (sin 50°) (sin $x°$):

$$9.5 \sin x° = 8.7 \sin 50°$$

∴

$$\sin x° = \frac{8.7 \sin 50°}{9.5}$$

Thus

$$x \approx 44.6 \text{ (correct to 1 decimal place)}$$

Or, using the 'solve' ability of some calculators:

We then dismiss the obtuse angle because x had to be smaller than 50. (Or, had we not noticed this from the side lengths, we would reject the obtuse angle as the angle sum of the triangle would exceed 180°)

Thus, as before, $x = 44.6$ (correct to 1 decimal place).

$$\text{solve}\left(\frac{9.5}{\sin(50)} = \frac{8.7}{\sin(x)}, x\right) | 0 \le x \le 180$$
$$\{x = 135.4496775, \ x = 44.55032253\}$$

The cosine rule

Again, consider a triangle ABC as shown below left for an acute angled triangle and below right for an obtuse angled triangle. However, in this case, we use the fact that for cosines:

$$\cos(180° - C) = -\cos C$$

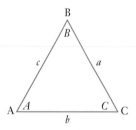 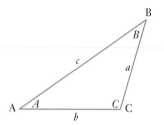

Again we draw the perpendicular from B to meet AC at D:

Again we draw the perpendicular from B to meet AC produced at D:

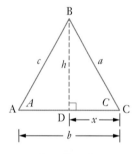

From △CBD: $a^2 = h^2 + x^2$ [1]

From △ABD: $c^2 = h^2 + (b - x)^2$

$= h^2 + (b - x)(b - x)$

i.e. $c^2 = h^2 + b^2 + x^2 - 2bx$

Using [1]: $c^2 = a^2 + b^2 - 2bx$ [2]

From △CBD: $\cos C = \dfrac{x}{a}$

∴ $x = a \cos C$ [3]

Using [2] and [3]:

$$c^2 = a^2 + b^2 - 2ab \cos C$$

From △CBD: $a^2 = h^2 + x^2$ [1]

From △ABD: $c^2 = h^2 + (b + x)^2$

$= h^2 + (b + x)(b + x)$

i.e. $c^2 = h^2 + b^2 + x^2 + 2bx$

Using [3]: $c^2 = a^2 + b^2 + 2bx$ [4]

From △CBD: $\cos(180° - C) = \dfrac{x}{a}$

∴ $x = -a \cos C$ [5]

Using [4] and [5]:

$$c^2 = a^2 + b^2 - 2ab \cos C$$

Thus for both the acute triangle and the obtuse triangle $c^2 = a^2 + b^2 - 2ab \cos C$.

This is the **cosine rule**:

$$c^2 = a^2 + b^2 - 2ab \cos C$$

Similarly $a^2 = b^2 + c^2 - 2bc \cos A$ and $b^2 = a^2 + c^2 - 2ac \cos B$

As was said with the sine rule, rather than learning the rule as a formula instead notice the pattern of what it is telling you:

The square of any side of a triangle is equal to the sum of the squares of the other two sides take away twice the product of the other two sides multiplied by the cosine of the angle between them.

EXAMPLE 6

Find the value of x for the triangle shown.

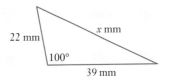

The cosine rule— angles and sides

Solution

By the cosine rule:

$$x^2 = 22^2 + 39^2 - 2\,(22)\,(39)\cos 100°$$

$$\approx 2302.98$$

$$x = 48 \text{ to the nearest integer.}$$

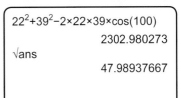

```
22²+39²−2×22×39×cos(100)
                  2302.980273
√ans
                  47.98937667
```

EXAMPLE 7

Find the value of x for the triangle shown.

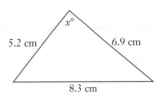

Solution

By the cosine rule:

$$8.3^2 = 5.2^2 + 6.9^2 - 2\,(5.2)\,(6.9)\cos x°$$

$$\cos x° = \frac{5.2^2 + 6.9^2 - 8.3^2}{2(5.2)(6.9)}$$

$$\approx 0.08027$$

$$x = 85 \text{ to the nearest integer.}$$

```
 5.2²+6.9²−8.3²
 ─────────────
   2×5.2×6.9
                  0.08026755853
cos⁻¹ (ans)
                  85.39605483
```

Note:

- If you prefer to use the solve facility on your calculator make sure you can obtain the same answers as those shown.

- With the cosine rule, when solving equations of the form cos $x = c$ we do not have to worry about there being two solutions in the range 0° to 180°. The cosine of an acute angle is positive whilst the cosine for an obtuse angle is negative. Thus an equation of the form cos $x = c$ does *not* have two solutions for x in the range 0° to 180°. If c is positive the one solution will be an acute angle and if c is negative it will be an obtuse angle.

EXAMPLE 8

The sketch on the right shows a system of three triangles with lengths and angles as indicated.

BAE is a straight line.

Find the length of CD.

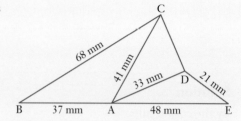

Thoughts

CD is one side of △ACD. In this triangle we know the lengths of AC and AD so if we knew the size of ∠CAD we could apply the cosine rule to find the length of CD. We can find the size of ∠CAD if we first find the size of ∠CAB and the size of ∠DAE.

Solution

For △ABC, applying the cosine rule:

$$68^2 = 41^2 + 37^2 - 2 \times 41 \times 37 \cos \angle BAC$$

$$\cos \angle BAC = \frac{41^2 + 37^2 - 68^2}{2 \times 41 \times 37}$$

∴ ∠BAC ≈ 121.3°

```
(41² + 37² – 68²) ÷ (2 × 41 × 37)
                        –0.5187870798
cos⁻¹ Ans
                          121.2509263
Ans → A
                          121.2509263
```

For △DAE, applying the cosine rule:

$$21^2 = 33^2 + 48^2 - 2 \times 33 \times 48 \cos \angle DAE$$

$$\cos \angle DAE = \frac{33^2 + 48^2 - 21^2}{2 \times 33 \times 48}$$

∴ ∠DAE ≈ 21.3°

```
(33² + 48² – 21²) ÷ (2 × 33 × 48)
                         0.9318181818
cos⁻¹ Ans
                          21.27996647
Ans → B
                          21.27996647
```

ISBN 9780170390262

For △CAD, applying the cosine rule:

$$CD^2 = 33^2 + 41^2 - 2 \times 33 \times 41 \cos \angle CAD$$

$$\approx 622.3$$

∴ $CD \approx 24.9$

CD is of length 25 mm, to the nearest millimetre.

```
33² + 41² – 2 × 33 × 41cos(180–A–B)
                          622.297976
√Ans
                          24.94590099
```

Notice from the calculator displays that the more accurate values for $\angle BAC$ and $\angle DAE$ were stored and later recalled for use, thus avoiding the risk of introducing unnecessary rounding errors.

Exercise 11C

The sine rule

Given that each of the following equations are formed by applying the sine rule to an acute angled triangle solve for x, giving your answer correct to one decimal place in each case.

1 $\dfrac{x}{\sin 50°} = \dfrac{7.3}{\sin 75°}$ **2** $\dfrac{x}{\sin 32°} = \dfrac{12.1}{\sin 78°}$ **3** $\dfrac{12.3}{\sin 60°} = \dfrac{x}{\sin 65°}$

4 $\dfrac{8.2}{\sin x°} = \dfrac{10}{\sin 85°}$ **5** $\dfrac{7.8}{\sin x°} = \dfrac{8.3}{\sin 50°}$ **6** $\dfrac{6.8}{\sin 50°} = \dfrac{7.2}{\sin x°}$

Find the value of x in each of the following.

7

8
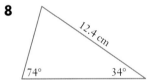

9

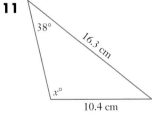

x is between 0 and 90.

10
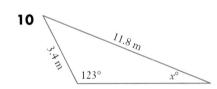

11

x is between 90 and 180.

12
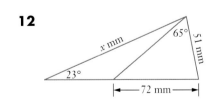

13

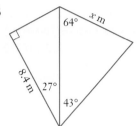

14

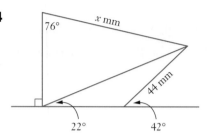

15 The diagram shows a pole AB with end A fixed on horizontal ground and the pole supported by a wire attached to end B and to a point C on the ground with AC = 420 centimetres.

The pole makes an angle of 50° with the ground and the wire makes an angle of 30° with the ground, as shown in the diagram.

Points A, B and C all lie in the same vertical plane.

Find the length of the pole giving your answer to the nearest centimetre.

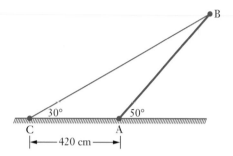

16 Rather than risking the direct shot over a lake a golfer prefers to take two shots to get to the green as shown in the diagram on the right.

How much further is this two shot route than the direct route?

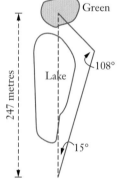

The cosine rule

Given that each of the following equations are formed by applying the cosine rule to an acute angled triangle solve for x in each case, giving your answers correct to one decimal place.

17 $x^2 = 7^2 + 8^2 - 2\,(7)\,(8)\cos 56°$

18 $x^2 = 3^2 + 2^2 - 2\,(3)\,(2)\cos 32°$

19 $3^2 = 5^2 + 7^2 - 2\,(5)\,(7)\cos x°$

20 $12^2 = 9^2 + 11^2 - 2\,(9)\,(11)\cos x°$

Find the value of x in each of the following.

21

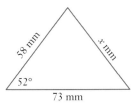

58 mm, x mm, 52°, 73 mm

22

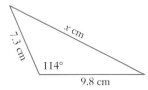

7.3 cm, x cm, 114°, 9.8 cm

23

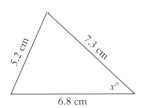

5.2 cm, 7.3 cm, $x°$, 6.8 cm

24

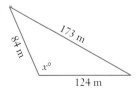

84 m, 173 m, $x°$, 124 m

25

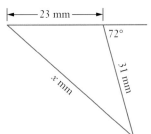

23 mm, 72°, x mm, 31 mm

26

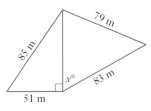

85 m, 79 m, $x°$, 51 m, 83 m

27

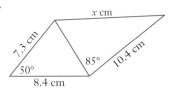

x cm, 7.3 cm, 85°, 10.4 cm, 50°, 8.4 cm

28

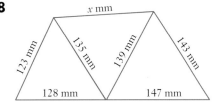

x mm, 123 mm, 135 mm, 139 mm, 143 mm, 128 mm, 147 mm

29 A boat travels 6.3 km due North and then turns 17° towards the West and travels a further 7.2 km. How far is it then from its initial position?

30 Jim and Toni leave the same point at the same time with Jim walking away at a speed of 1.4 m/s and Toni at a speed of 1.7 m/s, the two directions of travel making an angle of 50° with each other. If they both continue on these straight line paths how far are they apart after 8 seconds?

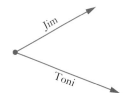

Jim, Toni

31 From location A, location B is 12.3 km away on a bearing of 070°.

From location A, location C is 7.2 km away on a bearing of 150°.

How far is B from C?

Miscellaneous

Find the value of x in each of the following.

32

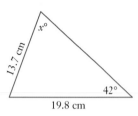

$x°$

13.7 cm

42°

19.8 cm

x is between 0 and 90.

33

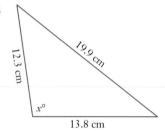

12.3 cm

19.9 cm

$x°$

13.8 cm

34

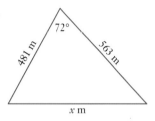

72°

481 m

563 m

x m

35

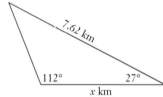

7.62 km

112°

27°

x km

36

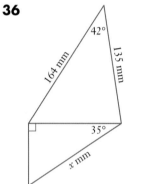

42°

164 mm

135 mm

35°

x mm

37

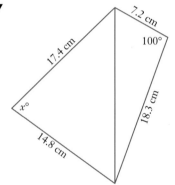

7.2 cm

100°

17.4 cm

$x°$

18.3 cm

14.8 cm

38

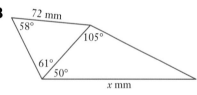

72 mm

58°

105°

61°

50°

x mm

39

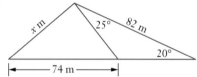

x m

25°

82 m

20°

74 m

40 The diagram on the right shows a mobile crane used to lift containers from ships and transfer them to waiting container trucks. If AB is of length 300 centimetres find the lengths of AC and BC.

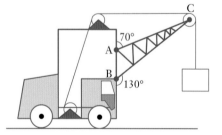

C

70°

A

B

130°

ISBN 9780170390262

41 A triangle has sides of length 12.7 cm, 11.9 cm and 17.8 cm. Find the size of the smallest angle of the triangle, giving your answer to the nearest degree.

42 A parallelogram has sides of length 3.7 cm and 6.8 cm and the acute angle between the sides is 48°. Find the lengths of the diagonals of the parallelogram.

43 The diagonals AC and BD, of parallelogram ABCD, intersect at E.

If ∠AED = 63° and the diagonals are of length 10.4 cm and 14.8 cm use the fact that the diagonals of a parallelogram bisect each other to determine the lengths of the sides of the parallelogram.

44 The tray of the tip truck shown on the right is tipped by the motor driving rod BC clockwise about B. As the tray tips, end C moves along the guide towards A.

If AB = 2 metres and BC = 1 metre find the size of ∠CAB when AC is

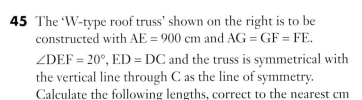

a   2.6 metres   **b**   2.1 metres.

45 The 'W-type roof truss' shown on the right is to be constructed with AE = 900 cm and AG = GF = FE.

∠DEF = 20°, ED = DC and the truss is symmetrical with the vertical line through C as the line of symmetry. Calculate the following lengths, correct to the nearest cm

a   CE   **b**   ED   **c**   DF   **d**   CF

46 Find, to the nearest millimetre, the distance between the tip of the 70 mm hour hand and the tip of the 90 mm minute hand of a clock at

a   5 o'clock

b   10 minutes past 5.

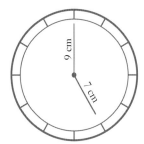

47 A coastal observation position is known to be 2.50 km from a lighthouse. The coastguard in the observation position is in radio and visual contact with a ship in distress at sea. If the coastguard looks towards the lighthouse and then towards the ship these two directions make an angle of 40° with each other. If the captain on the ship looks towards the observation position and then towards the lighthouse these two directions make an angle of 115° with each other. (The ship, the lighthouse and the observation position may all be assumed to be on the same horizontal level.)

How far is the ship from

a   the lighthouse?

b   the coastal observation position?

48 From a lighthouse, ship P is 7.3 km away on a bearing 070°.

A second ship Q is on a bearing 150° from P and 130° from the lighthouse.

a   How far is Q from P?

b   How far is Q from the lighthouse?

49 From a lighthouse, ship A is 15.2 km away on a bearing 030° and ship B is 12.1 km away on a bearing 100°. How far, and on what bearing, is B from A?

50 Ignoring any wastage needed for cutting, joining etc. what total length of steel would be needed to make twelve of the steel frameworks shown sketched on the right, rounding your answer up to the next ten metres.

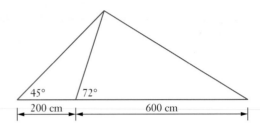

51 The diagram on the right shows the sketch made by a surveyor after taking measurements for a block of land ABCD.

Find the area and the perimeter of the block.

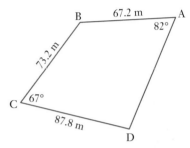

52 An engineering component consists of a rectangular metal plate with a triangular piece removed, as in the diagram below left.

The removed piece is cut away by a computer controlled machine that is programmed to cut a triangle with vertices at the distances and angles shown on the diagram below right.

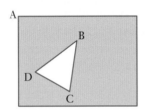 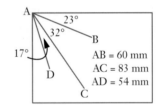

Find the area and the perimeter of the triangular piece that is removed.

53 The diagram on the right shows the sketch made by a surveyor after taking measurements for a block of land.

Find the area of the block.

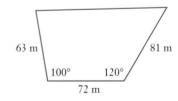

54 Make use of the cosine rule, and the rule for the area of a triangle given two sides and the included angle, to determine the area of a triangular block of land with sides of length 63 m, 22 m and 55 m and then check that your answer agrees with the following statement of the rule known as **Heron's rule**.

Area of a triangle with sides of length a, b and c is given by:

$$\text{Area} = \sqrt{s(s-a)(s-b)(s-c)} \quad \text{where } s = \frac{a+b+c}{2}.$$

ISBN 9780170390262

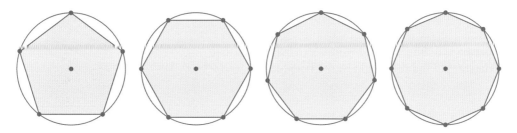
Miscellaneous exercise eleven

This miscellaneous exercise may include questions involving the work of this chapter, the work of any previous chapters, and the ideas mentioned in the Preliminary work section at the beginning of the book.

1 Find the equation of the straight line passing through (1, 1) and (4, 7).
Determine which of the points listed below lie on this line.
A(7, 15),   B(7, 13),   C(2, 2),   D(–1, 3),   E(6, 11).

2 Solve the following equations.

   a $3x = 51$ **b** $3x + 11 = 32$ **c** $2(3x + 2) – 5 = 11$

   d $5 – 2(3x + 2) = 13$ **e** $\dfrac{x}{5} = 7$ **f** $\dfrac{x}{5} + 3 = 7$

3 Use the midpoint of each interval to estimate a mean for the following distribution of fifty scores.

Score	$1 \to 5$	$6 \to 10$	$11 \to 15$	$16 \to 20$	$21 \to 25$	$26 \to 30$	$31 \to 35$
Frequency	8	14	9	7	6	4	2

In what interval does the median score lie?

4 In a test the 12 boys in a class scored a mean of 23.4 and the 16 girls in the class scored a mean of 24.1. Find the mean of the whole class of 28 students.

5 I think of a number, double it, add three, multiply the answer by three and then add on twice the number I first thought of. If my final answer is one hundred and forty five, what was the number I first thought of?

6 How long does it take an investment of $2500 to grow to $3220 in an account paying simple interest at the rate of 6.4% per annum?

7 A and B are two points on level ground, 13 metres apart. A vertical flagpole at B subtends an angle of 50° at the eye of a person standing at A and whose 'eye height' is 1.6 m (see diagram). Find the height of the flagpole.

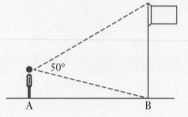

8 Ten scores are shown below in ascending order, lowest score on the left:

$$a - 1, \quad d, \quad 3d - 10, \quad 2d - 1, \quad a + 9, \quad c + 2, \quad c + 2, \quad 6e + 7, \quad c + 11, \quad 7b - 4.$$

The box plot for the ten scores is shown below:

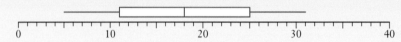

Find a, b, c, d and e and hence list the ten scores in ascending order.

9 In this chapter, we have developed the sine and cosine rules and a formula for the area of a triangle so that we can determine the area and unknown side lengths and angle sizes of triangles that are not right angled. Could these rules be applied to right angled triangles? What happens when these formulae are applied to right angled triangles? Investigate.

10 Two boats leave a harbour at the same time.

One travels due East at 7 km/hour and the other North-East at 5 km/hour.

How far are the boats apart 90 minutes later, to the nearest 100 metres?

11 A print shop commissions a sign maker to make and install an advertising sign.

The sign maker plans to suspend the sign from a framework as shown on the right.

The wall is vertical, DC is vertical and BF is horizontal. △BAC is equilateral, BA is of length 2 metres, ACF is a straight line and E is the midpoint of CF.

Find the length of

a   AF

b   BF

c   CD

d   CE

(Give answers in metres, correct to two decimal places if rounding is necessary.)

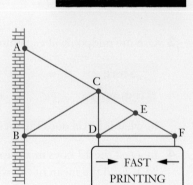

12 For the situation shown on the right how much shorter is the direct journey from A to C than the journey from A to C via B.

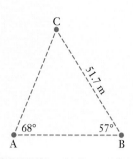

13 Electrical cabling is to be installed to connect three locations, P, Q and R whose relative positions are as shown in the diagram on the right.

Direct connection from P to R is not feasible so three possibilities are considered as shown below:

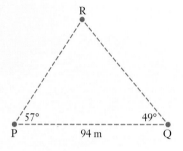

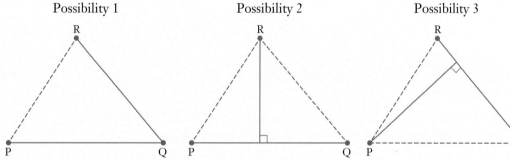

Find the total length of each of these giving each answer to the nearest metre.

14 To provide cover for nurses absent due to sickness and other reasons, a health authority maintains a 'pool' of nurses who are not attached to any particular ward or hospital but who can be directed to any particular area that is suffering a shortage. To assess how many nurses they need 'in pool' they collect information regarding how many nurses are absent on each day for one year. The information is shown in the following table:

No. of nurses absent	0 to 4	5 to 9	10 to 14	15 to 19	20 to 24	25 to 29	30 to 34
No. of days	56	137	97	46	18	9	2

Use the interval midpoints to calculate the mean number of nurses absent per day over this 365 day period and determine the standard deviation of the distribution.

If the authority decides to have n nurses in pool where n is given by:

n = the next integer after (mean + 0·5 × standard deviation),

how many do they decide to have in pool?

15 The graphs below show the 'population pyramids' of 2 countries, A and B.

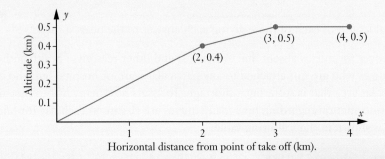

	% of Popl	MALE	FEMALE	% of Popl
AGE DISTRIBUTION COUNTRY A				
Age	% of Popl	MALE	FEMALE	% of Popl
65+	4.1			5.0
55-64	4.3			4.6
45-54	5.2			5.2
35-44	7.1			7.0
25-34	8.3			8.3
15-24	8.2			8.3
5-14	8.7			8.6
0-4	3.6			3.5

% of Popl	MALE	FEMALE	% of Popl	Age
AGE DISTRIBUTION COUNTRY B				
% of Popl	MALE	FEMALE	% of Popl	Age
2.5			1.8	65+
2.6			2.3	55-64
3.0			3.2	45-54
4.3			5.0	35-44
6.2			6.5	25-34
8.2			8.6	15-24
15.0			14.3	5-14
8.2			8.3	0-4

a In each pyramid the shaded bar at the bottom is smaller than the one above it? Suggest a reason why this might be.

b For which of the age ranges does A have more males than females?

c For which of the age ranges does B have more females than males?

d Country B has a population of 52 000 000. How many people aged 55 and over does the country have?

e Which of the two countries do you think is the more highly developed? Give reasons for your choice.

16 The diagram below shows the approximate path of an aircraft from take off to a point 4 km horizontally from take off drawn as three straight line sections.

y

Altitude (km)

0.5
0.4
0.3
0.2
0.1

(2, 0.4)   (3, 0.5)   (4, 0.5)

1   2   3   4   *x*

Horizontal distance from point of take off (km).

Find the equation of each of the three lines giving your answer in the form:

For *x* from 0 to 2: $y = ...$

For *x* from 2 to ...: $y = ...$

For *x* ...: $y = ...$

ISBN 9780170390262

12.

Simultaneous linear equations

- Introducing two variables
- Solving word problems
- Miscellaneous exercise twelve

Introducing two variables

The following example, and the solution that follows, appeared in chapter 7 where it was used to show how the introduction of an x allowed an equation to be built up and solved.

EXAMPLE 1

An amateur drama group hire a theatre for their production. They expect to sell all of the 1200 tickets, some at $10 and the rest at $7. The group requires the ticket sales to cover their $4150 production costs, to allow a donation of $4000 to be made to charity and to provide a profit of $1000 to aid future productions. If they are to exactly achieve this target, and their expectations regarding ticket sales are correct, how many of the total 1200 tickets should they charge $10 for and how many should they charge $7 for?

Solution

Let the number of $7 tickets be	x
These will give an income of	$7x$ dollars
The number of $10 tickets will then be	$(1200 - x)$
These will give an income of	$10(1200 - x)$ dollars
Thus	$7x + 10(1200 - x) = 4150 + 4000 + 1000$
which can be simplified to	$12\,000 - 3x = 9150$
Solving gives	$x = 950$

The group should sell 950 tickets at $7 each and 250 tickets at $10 each.

Instead of introducing a single variable, x, for the number of $7 tickets, we could introduce **two variables**, x and y, where x is the number of $7 tickets and y is the number of $10 tickets:

They expect to sell 1200 tickets	$\therefore$	$x + y = 1200$	[1]
The total revenue must be $9150	$\therefore$	$7x + 10y = 9150$	[2]

We now have **two equations** each involving the same two unknowns, x and y.

Two equations involving the same two variables can be solved together, *simultaneously*, to determine the two variables.

These 'simultaneous equations' as they are called can be solved by:

- using the simultaneous equation solving capability of some calculators,

- using the two equations 'against each other' to reduce to just *one* equation involving *one* variable, which can then be determined.

- using a graphical approach.

The following example is a repeat of the example on the previous page but the solution given below uses *two* variables and demonstrates the three methods of solution mentioned on the previous page.

EXAMPLE 2

Example 1 revisited

An amateur drama group hire a theatre for their production. They expect to sell all of the 1200 tickets, some at $10 and the rest at $7. The group require the ticket sales to cover their $4150 production costs, to allow a donation of $4000 to be made to charity and to provide a profit of $1000 to aid future productions. If they are to exactly achieve this target, and their expectations regarding ticket sales are correct, how many of the total 1200 tickets should they charge $10 for and how many should they charge $7 for?

Solution

Let the number of $7 tickets be		x	
and the number of $10 be		y	
They expect to sell 1200 tickets	$\therefore$	$x + y = 1200$	[1]
The total revenue must be $9150	$\therefore$	$7x + 10y = 9150$	[2]

1   Solve using the equation solving capability of some calculators.

The display on the right shows the two equations put into a calculator and the values

$$x = 950$$

and $y = 250$ displayed.

$$\begin{cases} x + y = 1200 \\ 7x + 10y = 9150 \end{cases} \Big|\, x, y$$
$$\{x=950,\ y=250\}$$

These are the values that 'fit' both the equation $x + y = 1200$ and $7x + 10y = 9150$.

Check: 1(950) +   1(250) = 1200 ✓
and   7(950) + 10(250) = 9150 ✓

The group should sell 950 tickets at $7 each and 250 tickets at $10 each.

Some calculators require you to input the equations in a different form.

The display below left, for example, shows the equations put into the calculator

by entering: 1   1   1200   for $1x + 1y = 1200$
and   7   10   9150   for $7x + 10y = 9150$

Below right shows the pair of values that satisfy these two equations.

anX+bnY=Cn
$$\begin{array}{c c c c} & a & b & c \\ 1 & \begin{bmatrix} 1 & 1 & 1200 \\ 7 & 10 & 9150 \end{bmatrix} \end{array}$$

$\longrightarrow$

anX+bnY=Cn
$$\begin{array}{c c} X & \begin{bmatrix} 950 \\ 250 \end{bmatrix} \\ Y & \end{array}$$

As before, the values that fit both equations are $x = 950$ and $y = 250$.

Note: In the equation $7x + 10y = 9150$, 7 and 10 are the **coefficients** of x and y.

ISBN 9780170390262

2   Solve using the two equations 'against each other' to reduce to just one equation involving one variable.

Substitution

Method 1: Substitution

We have the two equations:

$$x + y = 1200 \qquad [1]$$
$$7x + 10y = 9150 \qquad [2]$$

From equation [1] we obtain y in terms of x:

$$y = 1200 - x$$

Substitute this expression for y into [2]:

$$7x + 10(1200 - x) = 9150$$

Thus

$$7x + 12\,000 - 10x = 9150$$

i.e.

$$12\,000 - 3x = 9150$$

Add $3x$ to both sides to make the x term positive:

$$12\,000 = 9150 + 3x$$

Subtract 9150 from both sides to isolate $3x$:

$$2850 = 3x$$

Divide both sides by 3 to isolate x:

$$950 = x$$

i.e.

$$x = 950$$

From $y = 1200 - x$ it then follows that

$$y = 1200 - 950$$
$$= 250$$

The group should sell 950 tickets at $7 each and 250 tickets at $10 each.

Method 2: Elimination

In this method the strategy is to manipulate the equations until the coefficient of one of the variables is the same (except possibly for their sign) in both equations, and then to either add or subtract the two equations to eliminate that variable.

We have the two equations:

$$x + y = 1200 \qquad [1]$$
$$7x + 10y = 9150 \qquad [2]$$

$\times$ [1] by 10 so that it features $10y$: $\quad x + y = 1200 \quad \times 10 \to \quad 10x + 10y = 12\,000 \quad [3]$

Keep [2] unchanged: $\quad 7x + 10y = 9150 \quad\quad\quad \to \quad 7x + 10y = 9150 \quad [2]$

Equation [3] – equation [2]: $\quad 3x = 2850$

$\therefore$ $\quad x = 950$

From $x + y = 1200$ it then follows that $\quad y = 250$

The group should sell 950 tickets at $7 each and 250 tickets at $10 each.

Note

If you are required to be able to solve simultaneous equations without using the inbuilt features of some calculators, make sure you attempt some of the questions of the next exercise in that way.

---

3 **Solve using a graphical approach.**

The equation $x + y = 1200$ has many possible solutions, some of which are shown below:

x	0	1	10	15	33	98	117	900	950	1116
y	1200	1199	1190	1185	1167	1102	1083	300	250	84
$x + y$	1200	1200	1200	1200	1200	1200	1200	1200	1200	1200

The equation $7x + 10y = 9150$ has many possible solutions, some of which are shown below:

x	0	10	20	50	100	210	250	900	950	1120
y	915	908	901	880	845	768	740	285	250	131
$7x + 10y$	9150	9150	9150	9150	9150	9150	9150	9150	9150	9150

Notice that the above tables both include $x = 950$ and $y = 250$, the pair of values that fit both of the equations. Thus $x = 950$ and $y = 250$ are the solutions to the equations. However we could not be sure of being this lucky when we randomly list possible pairs for each of two equations. However we can use this idea to solve the equations graphically.

The graph of the equation $x + y = 1200$ will pass through *all* of the points whose x and y coordinates fit the equation $x + y = 1200$.

Similarly the graph of $7x + 10y = 9150$ will pass through *all* of the points whose x and y coordinates fit the equation $7x + 10y = 9150$.

Using a graphic calculator to plot both lines, the x and y coordinates of the point where the lines intersect will fit both equations (and of course with both equations being for straight lines there will be only one such point of intersection). The display on the right shows the equations, their graphs and the coordinates of the point of intersection.

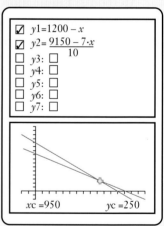

Thus solving the equations $\begin{cases} x + y = 1200 \\ 7x + 10y = 9150 \end{cases}$

gives $x = 950$ and $y = 250$.

The group should sell 950 tickets at $7 each and 250 tickets at $10 each.

- Some calculators require the equations to be input in the form '$y =$', as shown.
  Thus $x + y = 1200$ was entered as $y = 1200 - x$
  $7x + 10y = 9150$ was entered as $y = \dfrac{9150 - 7x}{10}$.

- You may need to adjust the viewing window of your calculator so that the point of intersection can be seen (though some calculators will state the coordinates of the point of intersection even if it is off screen).

Question: Which of these methods for solving simultaneous equations should you use?

Answer: Unless your teacher, or a particular question, requires you to use a particular method, use whichever method appeals to you most for that question, and with which you would expect to make least mistakes. **However, do make sure you know which methods you might be required to demonstrate in this unit and practise those.**

The method(s) you are likely to use need practice. The next two examples further demonstrate the technique of using two equations in two unknowns to give one equation in one unknown. This is not to suggest that this should be your chosen method for these questions but rather to further demonstrate its application and allow those readers who intend using other methods to check that they can obtain the same answers 'their way'.

Example 3 shows the substitution approach and **Example 4** the elimination approach.

EXAMPLE 3

a Solve $\begin{cases} 3x = 5y - 10 \\ x + y = 34 \end{cases}$

b Solve $\begin{cases} 0.3A + 0.1P = 161 \\ 5A - 3P = 1050 \end{cases}$

Solving simultaneous equations

Solution

a

$$3x = 5y - 10 \quad [1]$$
$$x + y = 34 \quad [2]$$

Simultaneous equations order activity

From [2]: $y = 34 - x$
Substitute into [1]: $3x = 5(34 - x) - 10$
Expand: $3x = 170 - 5x - 10$
Add $5x$ to both sides: $8x = 170 - 10$
i.e. $8x = 160$
Divide both sides by 8 $x = 20$
But $y = 34 - x$, thus $y = 34 - 20$
     $= 14$

Hence $x = 20$ and $y = 14$.

b

$$0.3A + 0.1P = 161 \quad [1]$$
$$5A - 3P = 1050 \quad [2]$$

$\times$ [1] by 10 to avoid decimals: $3A + P = 1610$
Thus $P = 1610 - 3A$
Substitute into [2]: $5A - 3(1610 - 3A) = 1050$
Expand: $5A - 4830 + 9A = 1050$
i.e. $14A - 4830 = 1050$
Add 4830 to both sides: $14A = 5880$
Divide both sides by 14: $A = 420$
But $P = 1610 - 3A$ and so $P = 1610 - 3 \times 420$
and so $P = 350$

Thus $A = 420$ and $P = 350$.

EXAMPLE 4

a Solve $\begin{cases} 3x + 2y = 11 \\ x + 2y = 1 \end{cases}$ b Solve $\begin{cases} 5x - 2y = 6 \\ 3x + 2y = 26 \end{cases}$ c Solve $\begin{cases} 2x + 3y = 12 \\ x + 4y = 11 \end{cases}$

Solution

a *Notice that both equations feature '+ 2y'. Taking one equation from the other will take '2y' from itself and thus eliminate one variable.*

	$3x + 2y = 11$	[1]
	$x + 2y = 1$	[2]
[1] – [2]:	$2x + 0y = 10$	
i.e.	$2x = 10$	
Thus	$x = 5$	
Substitute $x = 5$ into [2]:	$5 + 2y = 1$	
Take 5 from both sides:	$2y = -4$	
Hence:	$y = -2$	

Thus $x = 5$ and $y = -2$.

b *Notice that one equation features '– 2y' and the other features '+ 2y'. Adding the equations together will allow these to eliminate each other.*

	$5x - 2y = 6$	[1]
	$3x + 2y = 26$	[2]
[1] + [2]:	$8x + 0y = 32$	
i.e.	$8x = 32$	
Thus	$x = 4$	
Substitute $x = 4$ into [2]:	$12 + 2y = 26$	
Take 12 from both sides:	$2y = 14$	
Hence:	$y = 7$	

Thus $x = 4$ and $y = 7$.

c *If we leave the first equation unchanged but multiply the second equation by 2 we will have two equations each featuring '2x'. Taking one equation from the other will then eliminate one variable.*

$2x + 3y = 12$	[1]	$\rightarrow$		$2x + 3y = 12$
$x + 4y = 11$	[2]	$\times 2 \rightarrow$		$2x + 8y = 22$
Subtracting:				$-5y = -10$
Giving				$y = 2$
From [2]:				$x + 4(2) = 11$
				$x = 3$

Thus $x = 3$ and $y = 2$.

Solving word problems

In the next two examples, the two equations must first be determined from the information given and then the equations can be solved. Notice that in each example, the variables that are to be used are clearly introduced in the working.

Simultaneous
equations problems

EXAMPLE 5

Two numbers have a difference of 8 whilst three times the larger added to twice the smaller totals 59. Find the two numbers.

Solution

Let the smaller number be x and the larger number be y.

The two numbers have a difference of 8 $\qquad$ $\therefore$ $\qquad$ $y - x = 8$ $\qquad$ [1]

3 times the larger + 2 times the smaller = 59 $\qquad$ $\therefore$ $\qquad$ $3y + 2x = 59$ $\qquad$ [2]

Equations [1] and [2] can be solved to give $\qquad$ $x = 7$

and $\qquad$ $y = 15$

$$\begin{cases} y - x = 8 \\ 3y + 2x = 59 \end{cases} \Big| \, x, y$$
$$\{x=7, \, y=15\}$$

The two numbers are 7 and 15.

EXAMPLE 6

Every one of the 4000 tickets for a music concert at an entertainment centre is sold. Some of the tickets cost $28 each and the remainder cost $19 each. If the total revenue from the sale of the tickets is $83 200 find how many of the 4000 tickets cost $28 and how many cost $19.

Solution

Suppose x tickets cost $28 and y tickets cost $19.

There were 4000 tickets altogether $\qquad$ $\therefore$ $\qquad$ $x + y = 4000$ $\qquad$ [1]

The total revenue from the tickets sales was $83 200 $\qquad$ $\therefore$ $\qquad$ $28x + 19y = 83\,200$ $\qquad$ [2]

Equations [1] and [2] can be solved to give $\qquad$ $x = 800$

and $\qquad$ $y = 3200$

Thus 800 of the tickets cost $28 and 3200 of the tickets cost $19.

Notice that each of the two examples above finish with a clear statement of the answer. Example 5 does not end with $x = 7$ and $y = 15$. The question posed had no mention of x and y – we chose to introduce them to help us to solve the problem. The final answer should be expressed in the context of the question.

Hence example 5 ends with: *The two numbers are 7 and 15*

and example 6 with: *Thus 800 of the tickets cost $28 and 3200 of the tickets cost $19.*

Note: It is *not* the intention here to claim that these questions can only be solved by introducing two letters, building up two equations and solving them simultaneously. Questions like the previous example can be solved by introducing just one variable, as we saw at the start of this chapter.

Alternatively the solution could be 'reasoned through' as follows:

Selling all 4000 at $19 would have raised $4000 \times \$19 = \$76\,000$

However the ticket sales raised $83\,200$, i.e. $7200 'extra'. This extra must come from the extra $9 charged on some tickets.

$7200 is 800 lots of $9 so 800 seats were priced at $28 and 3200 at $19.

Yet another method would be to *guess* the number of $28 tickets there should be, *check* whether our guess works and then *improve* our guess, i.e. *guess*, *check* and *improve*.

Introducing two letters and solving the resulting pair of equations simultaneously can be very useful but other methods can be just as effective. In general, for each question you should choose the method you consider most appropriate for you to use to solve that question. However do use the next exercise to practise the techniques shown in this chapter.

iStock.com/Yuri_Arcurs

Exercise 12A

Use the graph shown on the right to solve the following pairs of equations simultaneously.

1 $y = x + 5$ and $y + 2x = 8$

2 $y = x - 4$ and $2y + x = 1$

3 $y = x - 4$ and $y + 2x = 8$

4 $y + 2x = 8$ and $2y + x = 1$

5 $y = x + 5$ and $2y + x = 1$

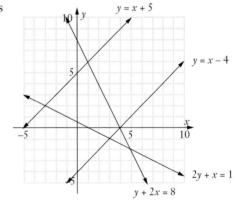

Solve the following pairs of equations.

6 $\begin{cases} 2x + y = 19 \\ x - y = 2 \end{cases}$

7 $\begin{cases} 3x + 2y = 17 \\ x + 2y = 11 \end{cases}$

8 $\begin{cases} 5x + 3y = 9 \\ 5x - y = 17 \end{cases}$

9 $\begin{cases} 3x + 2y = -1 \\ -x + 2y = 27 \end{cases}$

10 $\begin{cases} 3x - y = 16 \\ 2x + y = 29 \end{cases}$

11 $\begin{cases} 2x - 3y = 16 \\ x - 3y = 11 \end{cases}$

12 $\begin{cases} 3x + 5y = 47 \\ 3x + 2y = 26 \end{cases}$

13 $\begin{cases} -x + 7y = 3 \\ x - 3y = 1 \end{cases}$

14 $\begin{cases} 2x + 3y = 12 \\ 2x - y = -12 \end{cases}$

15 $\begin{cases} y = x - 3 \\ 2x + 3y = 11 \end{cases}$

16 $\begin{cases} 2x + y = 7 \\ 3x - 2y = 14 \end{cases}$

17 $\begin{cases} 3x - 2y = 6 \\ 2x - 3y = -1 \end{cases}$

18 $\begin{cases} y = 11 - 2x \\ 2x + 3y = 21 \end{cases}$

19 $\begin{cases} 3x - 5y = 6 \\ x = 2y + 1 \end{cases}$

20 $\begin{cases} 3A + 2B = 11 \\ 3A - 2B = 19 \end{cases}$

21 $\begin{cases} 2p - 3q = 2 \\ 4p + 2q + 1 = 29 \end{cases}$

22 $\begin{cases} 0.5x + 0.2y = 7 \\ 2x - 3y = -29 \end{cases}$

23 $\begin{cases} 2(x + 5) = 3y \\ x + 2y = 30 \end{cases}$

24 One day a baker bakes x white loaves and y wholemeal loaves.

   a The number of white loaves baked that day together with the number of wholemeal loaves baked that day totalled 600.

     Which of the following equations correctly expresses this information?

Equation 1:	Equation 2:	Equation 3:
$x - y = 600$	$y - x = 600$	$x + y = 600$

   b The number of white loaves baked that day exceeded the number of wholemeal loaves baked that day by 140. Which of the following equations correctly expresses this information?

Equation 4:	Equation 5:	Equation 6:
$x - y = 140$	$y - x = 140$	$x + y = 140$

   c Given that both of the statements from **a** and **b** are correct, solve your equations from parts **a** and **b** to determine the number of each type of loaf the baker baked that day.

25 At a dog show there were x people (each with two legs) and y dogs (each with four legs).

   a The total number of legs at the show that were either human legs or dog legs equalled 1758.

     Which of the following equations correctly expresses this information?

Equation 1:	Equation 2:	Equation 3:
$2x + 4y = 1758$	$4x + 2y = 1758$	$x + y = 1758$

   b If the number of dogs at the show is multiplied by 5 and the answer subtracted from the number of people at the show the number obtained is 403.

     Which of the following equations correctly expresses this information?

Equation 4:	Equation 5:	Equation 6:
$5x - y = 403$	$x - 5y = 403$	$5y - x = 403$

   c Given that both of the statements from **a** and **b** are correct solve your equations from parts **a** and **b** to determine the number obtained by adding the number of people at the show to the number of dogs at the show.

26 The smaller of two numbers is x and the larger is y.

a The larger of the two numbers is 5 more than the smaller.

Which of the following equations correctly expresses this information?

Equation 1:	Equation 2:	Equation 3:
$x - y = 5$	$y - x = 5$	$x + y = 5$

b Twice the smaller added to three times the larger equals 70.

Which of the following equations correctly expresses this information?

Equation 4:	Equation 5:	Equation 6:
$2x + 3y = 70$	$3x + 2y = 70$	$x + y = 70$

c Given that both of the statements from **a** and **b** are correct solve your equations from parts **a** and **b** to determine the two numbers.

27 Sally saves $1 and 50 cent coins by putting them into a piggy bank. When Sally opens the piggy bank she finds that it contains x $1 coins and y 50 cent coins.

a The piggy bank contained 46 coins altogether, all either $1 coins or 50 cent coins. Which of the following correctly expresses this information?

Equation 1:	Equation 2:	Equation 3:
$x + y = 46$	$xy = 46$	$x + 0.5y = 46$

b The total value of the coins in the piggy bank came to $32.

Which of the following correctly expresses this information?

Equation 4:	Equation 5:	Equation 6:
$2x + y = 32$	$x + 0.5y = 32$	$0.5x + y = 32$

c Solve your equations from parts **a** and **b** to determine how many of each type of coin the piggy bank contained.

28 A seamstress buys:

x metres of material A, costing $28 per metre,

and y metres of material B, costing $35 per metre.

a The seamstress buys a total of 23 metres of these two materials.

Which of the following correctly expresses this information?

Equation 1:	Equation 2:	Equation 3:
$x + y = 28$	$x + y = 35$	$x + y = 23$

b The two quantities cost the seamstress a total of $700 altogether.

Which of the following correctly expresses this information?

Equation 4:	Equation 5:	Equation 6:
$35x + 28y = 700$	$xy + 28 + 35 = 700$	$28x + 35y = 700$

c Solve your equations from parts **a** and **b** to determine what length of each material the seamstress bought.

ISBN 9780170390262

29 An investor invests \$$x$ in company X and \$$y$ in company Y.

a The investor invests a total of \$25 000 in these two companies.

Which of the following correctly expresses this information?

Equation 1:	Equation 2:	Equation 3:
$X + Y = 25\,000$	$x + y = \$25\,000$	$x + y = 25\,000$

b The investment in company X achieves a 4% loss in value and the investment in company Y achieves a 12% increase in value to make the total investment now worth \$25 120.

Which of the following correctly expresses this information?

Equation 4:	Equation 5:	Equation 6:
$-4x + 12y = 25\,120$	$0.04x + 0.12y = 25\,120$	$0.96x + 1.12y = 25\,120$

c Solve your equations from parts **a** and **b** to determine the amount the investor put into each company.

30 A rectangle has a length of x cm and a height of y cm.

a The perimeter of the rectangle is 70 cm.

Which of the following correctly expresses this information?

Equation 1:	Equation 2:	Equation 3:
$x + y = 35$	$xy = 70$	$x + y = 70$

b Three heights exceeds two lengths by 15 cm.

Which of the following correctly expresses this information?

Equation 4:	Equation 5:	Equation 6:
$3x - 2y = 15$	$3y - 2x = 15$	$3x + 2y = 15$

c Solve your equations from parts **a** and **b** to determine x and y and hence determine the area of the rectangle.

31 Entry into a particular event costs \$$x$ for each adult and \$$y$ for each child.

a For 16 adults and 7 children the total cost is \$256.

Which of the following equations correctly expresses this information?

Equation 1:	Equation 2:	Equation 3:
$16x + 7y = 265$	$16y + 7x = 256$	$16x + 7y = 256$

b For 20 adults and 11 children the total cost is \$338.

Which of the following equations correctly expresses this information?

Equation 4:	Equation 5:	Equation 6:
$20y + 11x = 338$	$20x + 11y = \$338$	$20x + 11y = 338$

c Solve your equations from parts **a** and **b** to determine the total cost of entry for five adults and three children.

32 Two numbers have a sum of 41 whilst three times the larger added to twice the smaller totals 106. By letting the smaller number be x and the larger be y express the given information as two equations and hence determine the two numbers.

33 Two numbers have a difference of eleven whilst five times the smaller exceeds twice the larger by seventeen. Find the two numbers.

34 A chemist is asked to make 100 mL of a particular medicine. This 100 mL should contain 20 g of a certain compound. The chemist has available two bottles of this medicine but neither is to the desired concentration. The solution in bottle A has 0.15 g per mL and the solution in bottle B has 0.4 g per mL. How many mL should the chemist use from each bottle to make the 100 mL of the desired concentration?
(Hint: Let the required mix consist of x mL from A and y mL from B.)

35 A person has $12 000 to invest in two companies, Acorp and Bcorp. The person invests $$x$ with Acorp and $$y$ with Bcorp. After one year each $1 invested with Acorp has grown to $1.12 and each $1 invested with Bcorp has grown to $1.05. The person's $12 000 has grown to $13 195.

a Write two equations involving x and y.

b Solve these equations to determine x and y.

36 The total amount received from the sale of 1500 tickets for a play is $13 800. Some of the tickets were sold for $12 each and the rest for $8 each. How many tickets were sold for $12 and how many for $8?

37 A coach hire company has 25 coaches. Some of these can carry 56 passengers each and the rest 35 passengers each. With all 25 coaches full 1211 passengers can be carried. How many of each size coach does the company have?

38 Two numbers are such that seven times the smaller exceeds three times the larger by one whilst twice the larger exceeds four times the smaller by four. Find the two numbers.

39 A stall at a school fete sold jars of jam, for $2.50 per jar, and jars of relish, for $2 per jar. In all they sold 78 jars of these two commodities, receiving a total of $179. How many jars of each commodity did they sell?

40 To start a new company a person borrows $120 000 from a bank. Under the terms of the loan the company will pay interest on this loan calculated at 14% per annum on part of the loan and 17% per annum on the remainder and does not have to repay any of the $120 000 capital until the second year. If the first year interest bill totalled $18 150 how much of the $120 000 was borrowed at 14% and how much at 17%?

41 A mathematics multiple choice test consisted of 25 questions. Candidates were awarded 4 marks for each correct answer, they lost 3 marks for each incorrect answer but there was no penalty for any questions that were left unattempted. David attempted 23 of the questions and scored 64 marks altogether. How many did he answer correctly?

Suppose instead that whilst 4 marks were still awarded for each correct answer and 3 were lost for each incorrect answer, 3 marks were also lost for each question not attempted. How many questions must be correctly answered for a mark of at least 50?

Miscellaneous exercise twelve

This miscellaneous exercise may include questions involving the work of this chapter, the work of any previous chapters, and the ideas mentioned in the Preliminary work section at the beginning of the book.

1 List all the errors evident in each of the following 'solutions'.

a

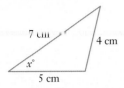

By the cosine rule:

$4^2 = 5^2 + 7^2 - 2 \times 5 \times 7 \cos x$

$16 = 25 + 49 - 70 \cos x$

$16 = 74 - 70 \cos x$

$16 = 4 \cos x$

$\cos x = 0.25$

$x = 75.522\,487\,81$

b

Using Area $= \dfrac{ab \sin C}{2}$

Area $= \dfrac{76 \times 72 \times \sin 64°}{2}$

$= 38 \times 36 \times \sin 32°$

≈ 725 cm

c In $\triangle ABC$, $\quad AB = 6.8$ cm,

$\quad\quad\quad\quad\quad BC = 7.1$ cm

and $\quad\quad\quad \angle BAC = 65°$.

To find the size of $\angle BCA$:

By the cosine rule:

$\dfrac{6.8}{\sin \angle BCA} = \dfrac{7.1}{\sin 65°}$

$\sin \angle BCA = \dfrac{7.1 \sin 65°}{6.8}$

≈ 0.9463

$\angle BCA = 71°$ or $109°$

to the nearest degree.

d

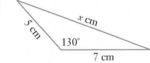

By the cosine rule:

$x^2 = 5^2 + 7^2 + 2 \times 5 \times 7 \cos 130°$

$= 25 + 49 + 70 \cos 130°$

$= 74 + 70 \cos 130°$

$= 144 \cos 130°$

≈ 92.5614

$x \approx 9.6$ cm

2 For each of the following classify the variable as one of the following four types:

| Nominal categorical | Ordinal categorical | Discrete numerical | Continuous numerical |

a Nationality.

b Height.

c Enthusiasm (High, Medium, Low).

d Number of people in family.

e Distance from home to work.

f Time for 400 metres.

g Number of people in a marathon.

h Gender.

i State of Australia a person's main residence is in.

j Number of press-ups completed in one minute.

3 Using only the graph shown on the right, determine the solutions to each of the following pairs of simultaneous equations.

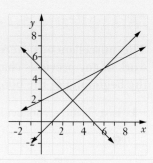

a $\begin{cases} y = x - 1 \\ y = 0.5x + 2 \end{cases}$

b $\begin{cases} y = x - 1 \\ y = -x + 5 \end{cases}$

c $\begin{cases} y = 0.5x + 2 \\ y = -x + 5 \end{cases}$

4 Solve the following equations.

a $5x - 7 = 11$

b $3(2x - 5) + 6 = 40 - x$

c $\dfrac{x + 3}{2} - 8 = 1$

d $\begin{cases} 5x - 3y = 46 \\ x + 2y = 17 \end{cases}$

e $\dfrac{x}{7} = \sin 30°$

f $\dfrac{7}{x} = \sin 30°$

5 The display on the right shows the lines

$x = 60$ $\qquad$ $y = 60$

$y = 2x - 60$ $\qquad$ $y = 0.5x + 30$

$y = -x + 60$ $\qquad$ $y = -2x + 30$

labelled A to F.

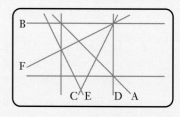

Allocate the correct rule to each line.

6 For the situation shown on the right how much shorter is the direct journey from A to C than the journey from A to C via B.

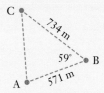

7 A particular 'family' of straight lines are related by the fact that they all have equations of the form $y = 3x + c$, each member having a different value for c.

What feature do the graphs of all members of this family have in common?

8 A particular 'family' of straight lines are related by the fact that they all have equations of the form $y = mx - 7$, each member having a different value for m.

What feature do the graphs of all members of this family have in common?

9 A particular 'family' of straight lines are related by the fact that they all have equations of the form $x + 2y = c$, each member having a different value for c.

What feature do the graphs of all members of this family have in common?

ISBN 9780170390262

10 From Lookout No. 1 a fire is spotted on a bearing 050°. From Lookout No. 2 the fire is seen on a bearing 020°. Lookout No. 2 is 10 km from Lookout No. 1 on a bearing 120°. Assuming that the fire and the two lookouts are all on the same horizontal level find how far the fire is from each lookout.

11 The road from A to B consists of two straight sections AC, length 2 km, and CB, length 3 km (see diagram). The bearing of C from A is 108° and the bearing of B from A is 132°.

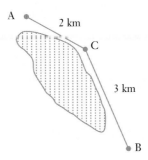

How much further is the road route from A to B than the straight line distance AB?

12 Two towns A and B are 60 km apart and are separated by a long road that can be assumed straight. A cyclist sets off from town B at 8 a.m. one morning and travels to town A in three stages, maintaining an approximately constant speed over each stage and resting for half an hour between stages. A delivery van sets off from town A, travels to town B at an approximately constant speed, stays in B for unloading etc, and then returns to A, again maintaining an approximately constant speed. The distance – time graph of this situation is shown below.

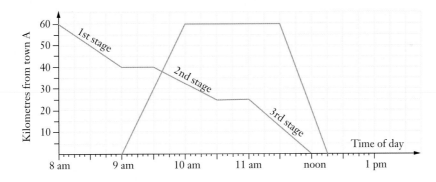

a Without doing any calculations it is possible to tell from the graph on which of the three stages the cyclist maintained the greatest average speed. Explain.

b What was the cyclist's average speed over each of the three stages?

c What was the delivery van's average speed from town A to town B?

d What was the delivery van's average speed from town B to town A?

e Without leaving town B earlier than it did what average speed would the delivery van have needed to travel back to A at if it was to arrive back at town A before the cyclist reached there?

13 An investor has $50 000 to invest for one year. She decides to put some of it in a secure deposit account and the rest in a more risky investment. At the end of the year the deposit account pays interest of 8%, the more risky investment pays 18% and the investor receives a total of $5600 in interest from these two sources. How much of the $50 000 had the investor put in the secure deposit account and how much in the more risky investment?

14 The investor of question 13 decides to re-invest her money for a second year. She adds the interest to her original $50 000 so she now has $55 600 to invest. She still opts to keep all of this money invested with either the secure deposit account or the more risky investment but she changes the balance of her portfolio. At the end of this second year her total interest is $4360 with the deposit account paying 10% and the more risky investment paying 7%. How much of the $55 600 had the investor put in the deposit account and how much in the more risky investment?

15 Triangle ABC has $a = 5$ cm, $b = 7$ cm and $c = 6$ cm.

  a Use the cosine rule to determine the size of $\angle C$.

  b Use $\dfrac{1}{2} ab \sin C$ to determine the area of $\triangle ABC$.

  c When given the lengths of the sides of a triangle an alternative way of determining the area is to use **Heron's 's' formula**:

  $$\text{Area of } \triangle ABC = \sqrt{s(s-a)(s-b)(s-c)} \qquad \text{where} \qquad s = \dfrac{a+b+c}{2}.$$

  Calculate the area of $\triangle ABC$ using this formula.

16 The times taken for some 12 year old students and some 14 year old students to run a particular distance were noted. Box plots of the times are shown below.

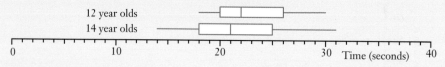

Each of the following statements are either incorrect or their correctness cannot be concluded from the boxplot.

  a Write a few sentences about each of the following statements.

  i *For the 12 year old results the box plot extends further to the right of the median than to the left. This shows there are more results involved to the right of the median than there are to the left.*

  ii *More 14 year olds were involved than 12 year olds.*

  iii *The interquartile range for the 14 year olds was much bigger than the interquartile range for the 12 year olds.*

  b Write at least five statements that in some way compare the times of the 12 year olds to those of the 14 year olds, with the first two of your five statements being completed versions of the following:

  i *The range of times for the 14 year olds (_____ seconds) exceeded the range of times for the 12 year olds (_____ seconds).*

  ii *The fastest 25% of the 14 year olds were _____ than the fastest of the 12 year olds.*

ISBN 9780170390262

13.

Standard scores and the normal distribution

- Standard scores
- Normal distribution
- Using a calculator
- In the old days: Using a book of tables
- Notation
- Quantiles
- Miscellaneous exercise thirteen

Test 1:
27

Kym sits a Mathematics test and achieves a mark of 27.
In the next test she scores 30. Has she improved?

Test 2:
30

Before answering this question we might first ask:
What was each test out of?

$\dfrac{27}{40}$

Suppose that test 1 was out of 40 and test 2 was out of
50. Can we now decide whether she has improved?

$\dfrac{30}{50}$

Before answering we may want to know if the tests were of
similar difficulty. What was the mean mark in each test?

Mean:
23

Suppose the mean in test 1 was 23 and in test 2 was 25.
Now can we judge whether her test 2 mark shows an
improvement?

Mean:
25

What if we also knew the standard deviation for
each test as well?

St dev:
5

Suppose the standard deviation in test 1 was 5 marks
and in test 2 was 10 marks

St dev:
10

Now can you suggest whether or not Kym's mark in
test 2 was an improvement on her mark in test 1?

13. Standard scores and the normal distribution ●●●●●●●●●●●●●

Standard scores

In the *Situation* on the previous page, did you consider expressing Kym's test scores in terms of the number of standard deviations each was from the mean? (An idea also encountered in one question of **Miscellaneous exercise three** earlier in this book.)

Expressing a score as a number of standard deviations above or below the mean is called **standardising** the score. We obtain the **standard score.**

$$\text{Standardised score} = \frac{\text{Raw score} - \text{mean}}{\text{standard deviation}}$$

EXAMPLE 1

Jennifer scores 23, 35 and 17 in tests A, B and C respectively. If the mean and standard deviation in each of these tests are as given below express each of Jennifer's test scores as standardised scores.

Test A:	mean	30	standard deviation	5
Test B:	mean	32	standard deviation	6
Test C:	mean	15	standard deviation	2.5

Solution

In Test A, Jennifer's standardised score is $\dfrac{23-30}{5}$ i.e. −1.4.

In Test B, Jennifer's standardised score is $\dfrac{35-32}{6}$ i.e. 0.5.

In Test C, Jennifer's standardised score is $\dfrac{17-15}{2.5}$ i.e. 0.8.

Exercise 13A

1 Express each of the following as a standard score.

  a A score of 65 in a test that had a mean of 60 and a standard deviation of 5.

  b A score of 72 in a test that had a mean of 55 and a standard deviation of 10.

  c A score of 50 in a test that had a mean of 58 and a standard deviation of 4.

  d A score of 60 in a test that had a mean of 58 and a standard deviation of 4.

  e A score of 58 in a test that had a mean of 64 and a standard deviation of 8.

2 SuMin scores 30, 50, 7 and 26 in tests A, B, C and D respectively. If the mean and standard deviation in each of these tests are as given below express each of SuMin's test scores as standardised scores.

Test A:	mean	20	standard deviation	4
Test B:	mean	60	standard deviation	10
Test C:	mean	6	standard deviation	0.8
Test D:	mean	25	standard deviation	5

3 All of the first year students on a particular technology course sat exams in the core subjects of Mathematics, Chemistry, Electronics and Computing. The exam results produced the following summary statistics:

Mathematics exam: mean mark   60   standard deviation   10.4

Chemistry exam: mean mark   72   standard deviation   7.2

Electronics exam: mean mark   48   standard deviation   14.6

Computing exam: mean mark   63   standard deviation   7.4

One student scored 56 in Mathematics, 74 in Chemistry, 39 in Electronics and 72 in Computing. Standardise each of these scores and rank the subjects for this student listing them from best to worst on the basis of these standard scores.

4 All year ten students in a particular region sat exams in Mathematics, English, Science and Social Studies. The exam results in these subjects produced the following means and standard deviations.

Mathematics: Mean   63   Standard deviation   14

English: Mean   64   Standard deviation   10

Science: Mean   72   Standard deviation   8

Social Studies: Mean   106   Standard deviation   22

One student achieved the following scores:

76 in Mathematics,   75 in English,   78 in Science,   104 in Social Studies.

Rank the four subjects in order for this student, highest standardised score first.

5 Jill and her boyfriend Jack sit the same maths exam, along with the 156 other candidates studying the course for which the exam formed a part of the assessment.

- The exam was marked out of 120.
- The mean mark for the entire 158 students was 65.2 and the standard deviation was 8.8.
- Jill scored 74 out of 120 and Jack scored 63 out of 120.

Complete the three incomplete responses from Jill shown below in the following conversation between her and her mother:

Jill (arriving home from school):   *'Hi Mum. How's your day been?'*

Jill's mum: *'Pretty good dear. How was yours? Did you get any marks back from the exams you did?'*

Jill: *'Yeah I got my maths mark.'*

Jill's mum: *'What did you get?'*

Jill, quoting her exam mark as a standard score replied:

'Well I got _____.'

Jill's mum: *'What! That sounds awful! What was the average?'*

Jill, again quoting standard scores: *'The mean was _____.'*

Jill's mum: *'What! What did Jack get?'*

Jill: *'Oh he got _____.'*

Jill's mum (who knew something about mathematics):

'Wait a minute. Are we talking standard scores here?'

Normal distribution

Suppose the diastolic blood pressure of a large number of adults was measured and the mean value was found to be 75 mm of mercury (mm of mercury being the units blood pressure is measured in). The data collected, if presented as a histogram, could well have a shape similar to the diagram shown on the right, i.e. a symmetrical distribution with many values close to the mean and the number of values decreasing as we move further from the mean.

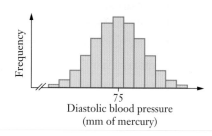

Fitting a smooth curve to the midpoints of the columns we obtain a '**bell shaped curve**' as shown on the right.

If we make many measurements of something that occurs naturally, for example the heights of many adult females, the weights of many domestic cats, the foot lengths of many adult males, etc., the histogram of the data often follows this sort of shape.

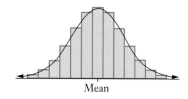

Data of this kind is said to be **normally distributed**. In **normal distributions**, approximately two thirds of the population lie within one standard deviation of the mean, 95% would lie within two standard deviations of the mean and almost all would lie within three standard deviation of the mean.

This is the 68%, 95%, 99.7% rule.

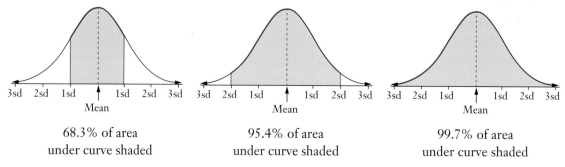

68.3% of area under curve shaded

95.4% of area under curve shaded

99.7% of area under curve shaded

In terms of probabilities, we could say that the probability of a randomly selected individual from a normally distributed population being within

- one standard deviation of the mean is 0.683,

- two standard deviations of the mean is 0.954,

- three standard deviations of the mean is 0.997.

> **Note**
>
> The normal distribution is also referred to as the Gaussian distribution, after the German mathematician Carl Gauss.

EXAMPLE 2

A box of breakfast cereal has 'contains 500 grams of breakfast cereal' printed on it. Suppose that in fact the weight of breakfast cereal contained in these boxes is normally distributed with a mean of 512 grams and a standard deviation of 8 grams. Determine the probability that a randomly chosen box of this cereal contains between 504 grams and 520 grams.

Solution

With a mean of 512 grams and a standard deviation of 8 grams:

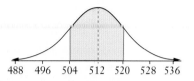

    504 grams is one standard deviation below the mean

and   520 grams is one standard deviation above the mean.

For normally distributed data the probability that a randomly chosen data point is within 1 standard deviation of the mean is, from the previous page, 0.683.

Thus the probability that a randomly chosen box of this cereal contains between 504 grams and 520 grams is 0.68.

The above example could be worked out using the '68' in the *68%, 95%, 99.7%* rule because the question involved numbers of standard deviations that this rule relates to. What would we have done if instead the question had asked for the probability of a randomly chosen box of the cereal containing less that 500 grams? In this case 500 grams is 1.5 standard deviations below the mean, a situation not covered by the 68%, 95% 99.7% rule. In this case we can use the ability of various calculators to determine such probabilities, as the next example (which is again based on the breakfast cereal situation of example 2) shows.

EXAMPLE 3

A box of breakfast cereal has 'contains 500 grams of breakfast cereal' printed on it. Suppose that in fact the weight of breakfast cereal contained in these boxes is normally distributed with a mean of 512 grams and a standard deviation of 8 grams.

a Determine the probability that a randomly chosen box of this cereal contains less than 500 grams.

b In a random sample of 100 boxes of this cereal approximately how many boxes should we expect to contain less than 500 g?

Solution

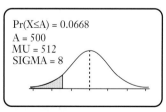

a For a randomly distributed set of values, with mean 512 and standard deviation 8, we require P(Randomly chosen value < 500).

Many calculators can display such information for normally distributed data.

The required probability is 0.0668.

The probability that a randomly chosen box of this cereal contains less than 500 grams is 0.0668.

b In any batch of boxes of this cereal we should expect that the proportion of them that contain less than 500 grams is about 0.07. Thus in a random sample of 100 boxes of this cereal we would expect approximately 7 boxes to contain less than 500 g.

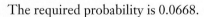

Using a calculator

The various calculators have different capabilities and routines with regard to displaying probabilities for normally distributed sets of data.

You will gain familiarity with the ability of *your* calculator in this regard in the next exercise.

In the old days: Using a book of tables

Prior to the ready availability of calculators with built in statistical routines for determining probabilities associated with normal distributions, these probabilities were determined using books of statistical tables.

These books give probabilities for just one normal distribution, the **standard normal distribution**. For this the random variable has a mean of 0 and a standard deviation of 1, as shown on the right.

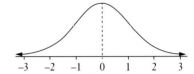

Normal distributions having means and standard deviations not equal to these standard values needed to be standardised. We encountered this idea of standardising data by expressing it as a number of standard deviations above or below the mean at the beginning of this chapter. Calling the original score an 'x score' and the standardised score a 'z score' we have:

$$z \text{ score } = \frac{x \text{ score } - \text{ mean of } x \text{ scores}}{\text{standard deviation of } x \text{ scores}}$$

Thus before the ready availability of sophisticated calculators, to answer the previous example which required us to determine the probability that from a normally distributed set of data, X, with mean 512 and standard deviation 8, a randomly selected item would have a value less than 500 we would have changed the 500 to a standard score:

$$\text{standard score } = \frac{500 - 512}{8}$$
$$= -1.5$$

(i.e. a score of 500 is 1.5 standard deviations below the mean)

and then used the table of probabilities for the standard normal distribution to determine the required probability.

z	0.00	0.01	0.02	0.03
−1.9	0.0287	0.0281	0.0274	0.0268
−1.8	0.0359	0.0351	0.0344	0.0336
−1.7	0.0446	0.0436	0.0427	0.0418
−1.6	0.0548	0.0537	0.0526	0.0516
−1.5	0.0668	0.0655	0.0643	0.0630
−1.4	0.0808	0.0793	0.0778	0.0764
−1.3	0.0968	0.0951	0.0934	0.0918

$$P(X < 500) = P(Z < -1.5)$$
$$= 0.0668$$

Thus, as before, the probability that a randomly chosen box of the cereal contains less than 500 grams is 0.0668.

Exercise 13B

The questions of this exercise refer to data sets involving normally distributed scores, X.

Using your calculator make sure that you can obtain each of the probabilities given in questions 1 to 8 below (correct to 4 decimal places), and each value of k in questions 9 to 17.

1

mean = 0
standard deviation = 1

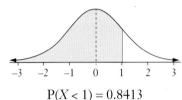

$P(X < 1) = 0.8413$

Can you also get 0.84 using the 68%, 95%, 99.7% rule?

2

mean = 0
standard deviation = 10

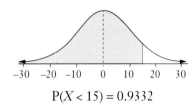

$P(X < 15) = 0.9332$

3

mean = 100
standard deviation = 25

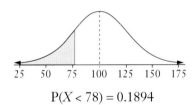

$P(X < 78) = 0.1894$

4

mean = 0
standard deviation = 1

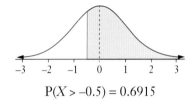

$P(X > -0.5) = 0.6915$

5

mean = 50
standard deviation = 10

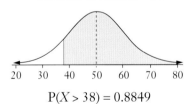

$P(X > 38) = 0.8849$

6

mean = 40
standard deviation = 4

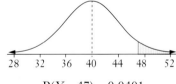

$P(X > 47) = 0.0401$

7

mean = 0
standard deviation = 1

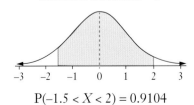

$P(-1.5 < X < 2) = 0.9104$

8

mean = 20
standard deviation = 4

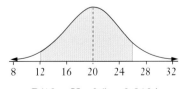

$P(12 < X < 26) = 0.9104$

9

mean = 0
standard deviation = 1

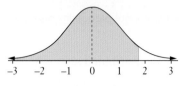

P(X < k) = 0.9573

∴ k = 1.72 (2 dp)

10

mean = 5
standard deviation = 1

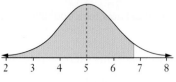

P(X < k) = 0.9671

∴ k = 6.84 (2 dp)

11

mean = 0
standard deviation = 1

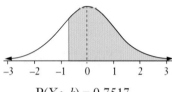

P(X > k) = 0.7517

∴ k = −0.68 (2 dp)

12

mean = 50
standard deviation = 5

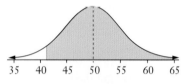

P(X > k) = 0.9656

∴ k = 40.9 (1 dp)

13

mean = 0
standard deviation = 1

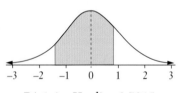

P(−1·4 < X < k) = 0.7215

∴ k = 0.85 (2 dp)

14

mean = 90
standard deviation = 2

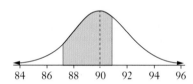

P(87.2 < X < k) = 0·5964

∴ k = 90.92 (2 dp)

15

mean = 40,
standard deviation = 15.
P(X < k) = 0.9850

∴ k = 72.55 (2 dp)

16

mean = 10,
standard deviation = 0.5.
P(X > k) = 0.0721

∴ k = 10.73 (2 dp)

17

mean = 0.1,
standard deviation = 0.01.
P(0.08 < X < k) = 0.3036

∴ k = 0.0955 (4 dp)

Notation

If we use X to represent the possible values of a normally distributed set of measurements having a mean μ and standard deviation σ (and hence variance σ^2) this is sometimes written:

$$X \sim N(\mu, \sigma^2).$$

'μ' is a Greek letter, mu, (pronounced myew) and σ is sigma so this is read as:

X is normally distributed with mean myew and standard deviation sigma.

EXAMPLE 4

If $X \sim N(63, 25)$ determine $P(X < 55)$.

Solution

X is normally distributed with a mean of 63 and a standard deviation of 5.

Using a calculator:

$P(X < 55) = 0.0548$

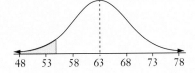

Using a tables book:

$P(X < 55) = P(Z < -1.6)$
$= 0.0548$

(Shown for interest only.)

EXAMPLE 5

Eight thousand two hundred and forty students were given an IQ test. The scores were normally distributed with a mean of 100 and a standard deviation of 16.

a Determine how many of the students, to the nearest ten, achieved a score in excess of 128.

b What were the minimum and maximum scores of the middle 60% of students on this test?

Solution

a Let X be the scores obtained in the test.
Thus $X \sim N(100, 16^2)$.
We require $P(X > 128)$.
Using a calculator, $P(X > 128) = 0.0401$.
Number scoring more than 128:
$$0.0401 \times 8240 \approx 330$$
Approximately 330 students achieved a score in excess of 128.

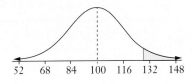

b If p is the lowest score achieved by the middle 60% then
$$P(X < p) = 0.2$$
i.e. $p = 86.53$
and if q is the highest score achieved by the middle 60% then
$$P(X < q) = 0.8$$
i.e. $q = 113.47$

(Some calculators can determine p and q more directly for this symmetrical situation.)

The lowest and highest scores achieved by the middle 60% are 86.5 and 113.5 respectively (to the nearest half mark).

EXAMPLE 6

If $X \sim N(40, 10^2)$ determine each of the following probabilities using the 68%, 95%, 99.7% rule, and not the statistical capability of your calculator.

a $P(30 < X < 50)$ **b** $P(20 < X < 60)$ **c** $P(40 < X < 60)$ **d** $P(X \leq 50)$

Solution

a 30 is one standard deviation below the mean and
50 is one standard deviation above the mean.
Thus $P(30 < X < 50) = 0.68$

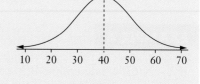

b $P(20 < X < 60) = 0.95$

c $P(40 < X < 60) = \dfrac{0.95}{2}$
$= 0.48$ (correct to 2 decimal places)

d $P(X \leq 50) = 0.5 + \dfrac{0.68}{2}$
$= 0.84$

Note that we make no distinction between $P(X \leq 50)$ and $P(X < 50)$. Including the line or not makes no difference to the area of the region.

EXAMPLE 7

Let us suppose that the time from Simon getting out of bed until his arrival at school is normally distributed with a mean of 55 minutes and a standard deviation of 5 minutes. Simon's arrival at school is classified as being late if it occurs after 9.10 am.

a One day Simon gets out of bed at 8.08 am. What is the probability of him arriving late?

b For a period of time Simon always gets out of bed at the same time but finds that he arrives late approximately 85% of the time! What time is he getting out of bed (to the nearest minute)?

Solution

a Let T minutes be the time from getting out of bed until
arrival at school.
Thus $T \sim N(55, 5^2)$.
Simon has 62 minutes to get to school before he is late.
We require: $P(T > 62)$
Calculator gives: $P(T > 62) = 0.0808$.
If Simon gets out of bed at 8.08 am the probability of him arriving late is 0.0808.

b The time that Simon is allowing himself to get to school is
causing him to be late approximately 85% of the time.
We require t for which $P(T > t) = 0.85$.
Calculator gives: $t \approx 49.8$
Thus Simon is allowing approximately 50 minutes to get to
school and for 85% of the days the journey takes longer than this, causing him to be late 85% of the time. Simon is getting out of bed at 8.20 am.

Quantiles

Quantiles are the values which a particular proportion of the distribution falls below.

Thus if 0.7 (70%) of the distribution is below 55, then 55 is the 0.7 quantile.

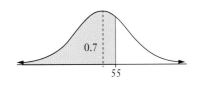

Alternatively we can refer to 55 as being the 70th **percentile**.

Note

- We are already accustomed to referring to the 0.25 quantile as the first, or lower, **quartile** and the 0.75 quantile as the third, or upper, quartile.

- If the quartiles divide a distribution in to four equal parts and the percentiles divide the distribution into 100 equal parts what might deciles and quintiles do?

EXAMPLE 8

If $X \sim N(20, 3^2)$ determine

a   the 0.5 quantile

b   the 0.82 quantile

c   the 24th percentile

d   the 62nd percentile

Solution

a
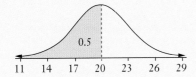

By inspection:
The 0.5 quantile is 20.

b

Using a calculator:
The 0.82 quantile is 22.7.

c
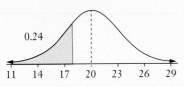

Using a calculator:
The 24th percentile is 17.9.

d
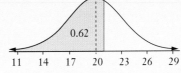

Using a calculator:
The 62nd percentile is 20.9.

Exercise 13C

1 The random variable, X, is normally distributed with a mean of 12 and a standard deviation of 2, i.e. $X \sim N(12, 2^2)$. Determine $P(X \geq 13.5)$.

2 The random variable, X, is normally distributed with a mean of 240 and a variance of 400, i.e. $X \sim N(240, 20^2)$. Determine $P(218 < X < 255)$.

3 $X \sim N(62, 64)$, i.e. X, is normally distributed with a mean of 62 and a standard deviation of 8. Given that $P(X > k) = 0.8238$ determine k.

4 If $X \sim N(0, 1)$ determine

   a   the 0.72 quantile,   **b**   the 0.26 quantile,

   c   the 89th percentile,   **d**   the 23rd percentile.

5 If $X \sim N(20, 3^2)$ determine

   a   the 0.44 quantile,   **b**   the 0.74 quantile,

   c   the 33rd percentile,   **d**   the 85th percentile.

6 Using the 68%, 95%, 99.7% rule, and *not* the statistical capability of your calculator, determine the following probabilities.

   a $P(-1 < X < 1), X \sim N(0, 1^2)$.   **b** $P(-2 < X < 2), X \sim N(0, 1^2)$.

   c $P(-3 < X < 3), X \sim N(0, 1^2)$.   **d** $P(8 < X < 32), X \sim N(20, 6^2)$.

   e $P(4 < X < 16), X \sim N(10, 2^2)$.   **f** $P(0 < X < 1), X \sim N(0, 1^2)$.

   g $P(X < 1), X \sim N(0, 1^2)$.   **h** $P(X > 1), X \sim N(0, 1^2)$.

   i $P(X < 5), X \sim N(0, 5^2)$.   **j** $P(X > 70), X \sim N(60, 10^2)$.

7 Let us suppose that the duration of pregnancy, for a naturally delivered human baby, is a normally distributed variable with a mean of 280 days and a standard deviation of 10 days.

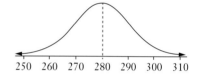

Using the 68%, 95%, 99.7% rule, and *not* the statistical capability of your calculator, determine estimates for the percentage of human pregnancies, for naturally delivered babies, that:

   a   are between 250 days and 310 days,

   b   exceed 290 days,

   c   are between 260 days and 270 days.

8 A machine produces components whose weights are normally distributed with a mean of 500 g and standard deviation of 5 g. According to the 68%, 95%, 99.7% rule, what percentage of the components will have:

   a   a weight of less than 495 g?

   b   a weight of less than 490 g?

9 A box of breakfast cereal has 'contains 300 grams of breakfast cereal' printed on it. Suppose that in fact the weight of breakfast cereal contained in these boxes is normally distributed with a mean of 310 grams and a standard deviation of 4 grams. Determine the probability that a randomly chosen box of this cereal contains:

   a   more than 312 grams of breakfast cereal,

   b   less than 300 grams of breakfast cereal.

ISBN 9780170390262

10 The lengths of adult male lizards of a particular species are thought to be normally distributed with a mean of 17.5 cm and a standard deviation of 2.5 cm.

Determine the probability that a randomly chosen adult male lizard of this species will have a length

a less than 17.5 cm

b between 15 cm and 17.5 cm.

11 The scaled scores in a national mathematics test are normally distributed with a mean of 69 and a standard deviation of 12.

What is the probability that a randomly selected candidate who sat this test has a scaled score of

a more than 75?

b between 66 and 75?

c less than 45?

12 The heights of fully grown plants of a certain species are normally distributed with a mean of 30 cm and a standard deviation of 4 cm. If 100 fully grown plants of this species are randomly selected approximately how many would you expect to be:

a taller than 35 cm?

b shorter than 25 cm?

c between 25 cm and 30 cm in height?

13 Let us suppose that 44 mg is 110% of the recommended daily intake of a particular vitamin and that a 110 mL container of fruit juice contains approximately 44 mg of this vitamin. If in fact the weight of the vitamin in the 110 mL containers of the fruit juice is normally distributed with mean 44 mg and standard deviation 2.5 mg, determine the probability that a randomly chosen 110 mL container of this fruit juice contains less than the recommended daily intake of the vitamin.

14 Five thousand five hundred and forty-two students sat a particular leaving exam. The scores obtained were normally distributed with a mean of 62 and a standard deviation of 12.5.

a Distinction certificates were awarded to students who gained a mark of 80 or more. How many students gained distinction certificates?

b A mark of less than 40 was regarded as a fail. How many of the students failed?

15 Let us suppose that the heights of the adults of a particular country are normally distributed with a mean of 1.75 m and a standard deviation of 10 cm. A car manufacturer wishes to design a new car with the space allowed for the driver, and the 'travel' on the drivers seat, suitable for every adult in the population except the tallest 5% of the adult population and the shortest 5% of the adult population. What is the height of the shortest driver and the tallest driver that the manufacturer is attempting to allow for? (Answer to nearest half centimetre.)

16 The marks achieved in a particular exam are normally distributed with a mean of 64 and a standard deviation of 12.

Grades are to be awarded as follows:

Top 12% of candidates: Grade A

Next 25% of candidates: Grade B

Next 40% of candidates: Grade C

Next 15% of candidates: Grade D

Remainder of candidates: Grade F

Determine the marks that form the A/B, B/C, C/D, and D/F grade boundaries, giving your answers correct to the nearest whole number.

17 Let us suppose that the time, in minutes, from Monica leaving home until she arrives at work is a normally distributed random variable with a mean of 45 and a standard deviation of 5. Monica's arrival at work is classified as late if it occurs after 8.30 am.

a One day Monica leaves home at 7.40 am. What is the probability of her arriving late?

b For a period of time Monica leaves home at the same time each day. During this period she finds that she arrives late approximately 8% of the time. What time is she leaving home (to the nearest minute)?

c What is the latest time (involving whole minutes) that Monica should leave home each day if she wishes to cut her late arrivals to less than 1%?

18 The annual rainfall in an area in the south west of Western Australia is normally distributed with a mean of 1200 mm and a standard deviation of 200 mm.

According to this model, and assuming the situation does not change, in every one hundred years how many years would you expect the annual rainfall to be

a less than 800 mm?

b more than 1500 mm?

c between 800 mm and 1500 mm?

(Challenge.) Given that a year has an annual rainfall of more than 1300 mm what is the probability that the rainfall for the year is less than 1500 mm?

19 The weight of each apple harvested from a particular orchard determines where the apple will be sent:

If   weight of apple ≥ 250 g   send to premium outlet

   150 g < weight of apple < 250 g   send to normal market

              weight of apple ≤ 150 g   send for juicing.

The weights of the apples are normally distributed with mean 180 g and standard deviation 40 g.

a In a random sample of 1000 apples how many would you expect to go to the premium outlet?

b (Challenge.) Given that an apple does not go to the premium outlet what is the probability that it is sent for juicing?

ISBN 9780170390262

Miscellaneous exercise thirteen

This miscellaneous exercise may include questions involving the work of this chapter, the work of any previous chapters, and the ideas mentioned in the Preliminary work section at the beginning of the book.

1 A particular straight line with a gradient of m and cutting the y-axis at the point with coordinates $(0, c)$ has equation $y = mx + c$.

   a The line passes through the point $(3, 4)$.

      Which of the following equations must be true?

Equation 1:	Equation 2:	Equation 3:
$y = 3x + 4$	$3 = 4m + c$	$4 = 3m + c$

   b The line also passes through the point $(8, 19)$.

      Which of the following equations must be true?

Equation 4:	Equation 5:	Equation 6:
$8 = 19m + c$	$19 = 8m + c$	$y = 8x + 19$

   c Solve your equations from parts **a** and **b** to determine the equation of the straight line.

2 Find the mean, median, mode and range for the set of scores shown in the dot frequency graph.

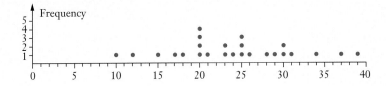

3 One hot dog and three burgers cost $13.30. Four hot dogs and four burgers cost $25.20. What would be the cost of four hot dogs and six burgers? (Assume that the individual cost of a hot dog and of a burger remain unchanged throughout.)

4 A rectangle is of length x cm and width y cm.

   Four times the width exceeds one length by 4 cm.

   The perimeter of the rectangle is 42 cm.

   Find the values of x and y and hence determine the area of the rectangle.

5 In support of a primary school fete a parent makes lots of high quality home made chocolates and lollies. The organisers decide to sell bags containing 20 chocolates for $6.00 and bags containing 20 lollies for $4.00. They also wish to sell a mixed bag of 20 chocolates and lollies for $4.80. How many chocolates and how many lollies should each of these bags hold for the price of $4.80 to be consistent with the prices of the other bags?

6 A road construction company charges 125 million dollars for constructing a 20 km stretch of highway and 245 million dollars for constructing 40 km of similar highway. Based on these costs and assuming a linear relationship exists between the total cost and the length of road constructed, determine the cost of constructing

a 25 km of similar highway

b 52 km of similar highway.

7 a Three consecutive integers have a sum of 504. Find the integers.

b Three consecutive even integers have a sum of 504. Find the integers.

8 Jack walks 2.4 km on a bearing 060° followed by 4.4 km on a bearing 190°. On what bearing and for what distance should he now walk to return directly to his starting point?

9 From a point A, level with the base of a monument, the angle of elevation of the topmost point of the monument is 35°.

From point B, also at ground level but 30 metres closer to the monument, the same point has an angle of elevation of 60°.

Find how high the topmost point is above ground level. (Give your answer correct to the nearest metre.)

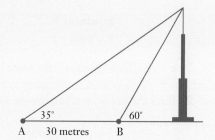

10 When a company sells x units of a particular product the revenue raised, $\$R$, is given by one of the lines on the graph shown and the other line shows $\$C$, the cost of producing these x units.

a Which of the two lines, I or II, is likely to be the revenue line and which the cost line? (Explain your answer.)

b What does the graph suggest is the value of x for 'break even'? (i.e. revenue raised = cost of production.)

c Suggest equations for each of the two lines shown.

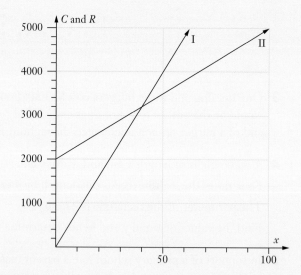

11 A yacht travels from location A to location B by tacking, as shown in the diagram.

The direct distance from A to B is 580 metres. The path the yacht takes causes it to cross the direct line from A to B at a point C where the distance from A to C is 65% of the distance from A to B.

How far did the yacht travel altogether in its journey from A to B?

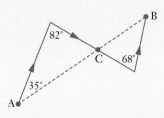

ISBN 9780170390262

12 A real estate agent wants to quote an average house price for a particular region. He obtains the following information about the sale prices of the 29 most recent sales in the area:

Sale price	$400 000 or less	$400 001 → $500 000	$500 001 → $600 000	$600 001 → $700 000
No. of sales	1	2	3	4

Sale price	$700 001 → $800 000	$800 001 → $900 000	$900 001 → $1 000 000	Over $1 000 000
No. of sales	5	5	3	6

What should he quote as an average price? Justify your answer and include mention of any issues you consider relevant.

13 An ornate window is to consist of two triangular pieces of glass placed in an aluminium frame as shown in the diagram on the right. Neglecting the thickness of the frame, find:

a   the total length of the aluminium

b   the total area of glass.

14 Clearly showing your use of trigonometry, determine to the nearest millimetre the distance between the tip of the 205 mm hour hand of a clock and the tip of the 312 mm minute hand of the clock at twenty minutes to four.

15 Twenty-four students sat a test and the scores they obtained were as follows:

Initials	PA	CB	JB	CC	JD	KD	LF	LJ	MJ	EK	IM	PN
Score	35	19	47	25	39	30	9	34	41	33	39	29

Initials	RN	PP	AR	TR	VR	AS	PS	TS	BV	PV	IW	RZ
Score	26	41	17	33	43	35	28	33	26	37	12	30

Grades of A, B, C, D and F are awarded according to the following rules:

Score $\geq$ (mean + 1.5 × standard deviation):   Grade A

(mean + 0.5 × standard deviation) $\leq$ Score < (mean + 1.5 × standard deviation):   Grade B

(mean − 0.5 × standard deviation) $\leq$ Score < (mean + 0.5 × standard deviation):   Grade C

(mean − 1.5 × standard deviation) $\leq$ Score < (mean − 0.5 × standard deviation):   Grade D

Score < (mean − 1.5 × standard deviation):   Grade F

Assign grades to each of the 24 students according to these rules.

16 Points A, B and C all lie on horizontal ground. A vertical tower DA has its base at A and is of height 15 metres. C lies due North of A and B is due East of A. The angle of elevation of D is 20° from C and 30° from B (see diagram).

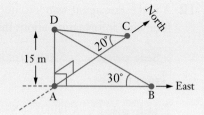

Calculate **a** how far C is from A

b how far B is from A

c how far C is from B

d the bearing of C from B.

17 The tank in a water irrigation system holds 80 000 litres of water.

The tank is initially three quarters full and each day, from 6am to noon, water flows from the tank at the rate of 1000 litres per hour.

This occurs for 6 days with no water entering the tank.

On day 7 rain is forecast so the system is switched off for days 7 and 8. This rain means that not only is no water taken off during these two days but instead, on day 7, from 6 am to 6 pm, 24 000 litres flows in. Unfortunately no water flows in on day 8.

The usual 6 am to noon daily outflow then recommences for days 9, 10, 11 and 12 and during these four days no more rain falls.

Sketch a piecewise graph showing the amount of water in the tank for these 12 days.

18 The 5347 candidates who sat a national mathematics test scored marks that had a mean of 127 and a standard deviation of 17.

- Distinction certificates are to be awarded to the top 15% of students.

- Participation certificates are to be awarded to students scoring less than 115. Candidates scoring between the above categories receive other certificates.

- The top 1% of students were awarded prizes.

Modelling the results as a normal distribution with mean 127 and standard deviation 17, determine each of the following:

a The lowest mark, rounded to the next whole mark *down*, that would achieve a distinction certificate.

b The number of students, to the nearest 10 students, who would receive a participation certificate.

c The lowest mark, rounded to the next whole number *up*, that would achieve a prize.

d The lowest and highest marks, each rounded to one decimal place, achieved by the middle 20% of students.

19 (Challenge) A river runs East to West. A tree stands on the edge of one bank and from a point C, on the opposite bank and due South of the tree, the angle of elevation of the top of the tree is 28°. From point D, 65 m due East of C the angle of elevation of the top of the tree is 20°.

Calculate **a** the height of the tree,

b the width of the river at C.

ANSWERS

Note:

- For questions that do not stipulate a specific level of rounding the answers given here have been rounded to a level considered appropriate for the question.

- If a question asks for an answer to be given 'to the nearest centimetre' it does not necessarily have to be given 'in centimetres'. In such a situation an answer of 174.9256 metres could be written as 174.93 m or as 17 493 cm, both answers being to the nearest centimetre.

Exercise 1A PAGE 7

1	Nominal	**2**	Nominal
3	Ordinal	**4**	Ordinal
5	Nominal	**6**	Ordinal
7	Ordinal	**8**	Nominal
9	Nominal	**10**	Ordinal
11	Nominal	**12**	Nominal

13

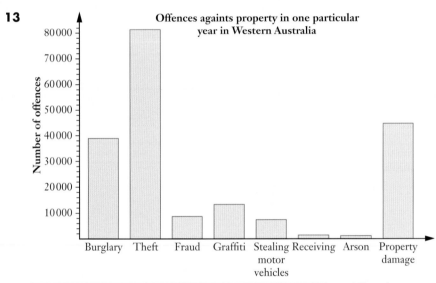

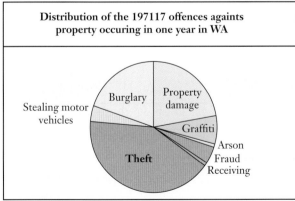

The advantage of the pie chart is that it shows how the total number of offences are divided up – i.e. the numbers of each type of offence as a proportion of the whole is shown. Whilst the pie chart also allows the numbers of each offence to be compared, if two categories are close in number it could be difficult to determine which has the greater number.

The bar chart on the other hand allows for very good comparison between categories but the proportion that each category is of the whole is not so evident.

Both the pie chart and the bar graph shown fail to show the accurate numbers given in the question but do allow overall comparisons to be made. However this weakness could be overcome if it was felt to be significant by including the accurate figures, with each category title in the pie graph, or at the top of each bar in the bar graph.

ISBN 9780170390262

14

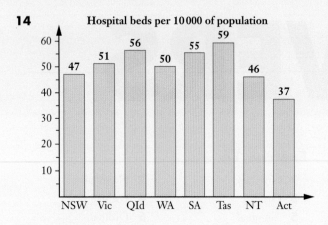

Hospital beds per 10 000 of population

Values: NSW 47, Vic 51, QId 56, WA 50, SA 55, Tas 59, NT 46, Act 37

Exercise 1B PAGE 8

1 Discrete	**7** Discrete
2 Continuous	**8** Continuous
3 Discrete	**9** Discrete
4 Continuous	**10** Continuous
5 Continuous	**11** Continuous
6 Continuous	**12** Continuous

Exercise 1C PAGE 13

1, 2, 3 Compare your choice with those of others in your class and discuss with your teacher.

4 Compare and discuss your sketches with those of others in your class. Discuss the reasonableness of your sketches and their sketches.

5

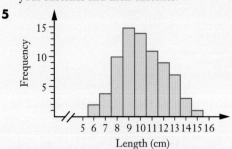

6

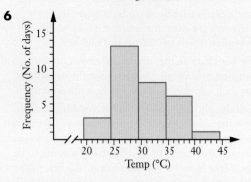

7

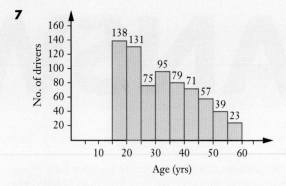

Values: 138, 131, 75, 95, 79, 71, 57, 39, 23

8

Time (secs)	30– 39	40– 49	50– 59	60– 69	70– 79	80– 89	90– 99
Frequency	2	32	53	39	14	7	3

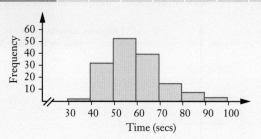

9

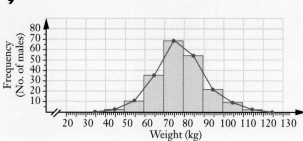

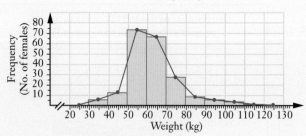

ISBN 9780170390262

10

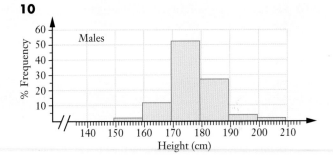

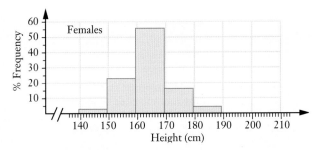

Miscellaneous exercise one   PAGE 17

1 **a** False   **b** False   **c** False
   d False   **e** False   **f** False

2 **a** Nominal categorical
   b Continuous numerical
   c Discrete numerical
   d Nominal categorical
   e Ordinal categorical
   f Nominal categorical
   g Continuous numerical
   h Discrete numerical

3 **a** 130   **b** 2.1   **c** 29.3

4 **a** 22   **b** 21   **c** 13.5

5 **a** 3   **b** 13 and 17   **c** 22

6 7.5 litres of water should be added to 25 millilitres of concentrate.

7 **a** mean 37.6,   median 38,   mode 40,   range 7
   b mean 102,   median 107,   mode 131,   range 63
   c mean 17.2,   median 17.5, mode 18,   range 4

8 **a** $15x - 6$ **b** $28 - 8x$
   c $-6x - 21$ **d** $8 - 16x$
   e $10p - 35$ **f** $6h - 15$
   g $7 + 6x$ **h** $-2x - 11$
   i $-4x + 15$ **j** $7 + 8q$
   k $7w + 22$ **l** $6p$

9 The framework requires 37 metres of steel (rounded up to the next whole metre).

10 **a** Frequency: 7, 18, 15, 11, 7, 3, 6, 8, 8, 5.
   b

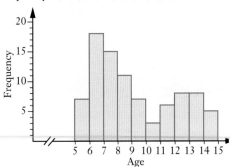

Exercise 2A   PAGE 25

1 mean 3.5 (1dp), median 4, mode 5, range 5.

2 mean 5.7 (1dp), median 6, mode 6, range 10.

3 mean 17.6 (1dp), median 18, mode 18, range 5.

4 mean 101.9, median 102, mode 101 and 103, range 7.

5 mean 6.3 (1dp), median 7, mode 9, range 9.

6 mean 25.2, median 28, mode 28 and 32, range 36.

7 **a** 118   **b** 164   **c** 139   **d** 139.8

8 **a** 48   **b** 94   **c** 77   **d** 76.4

9 A little less than two.

10 mean 6.3, median 6.5, mode 7.

11 Assuming the teacher is older than 17.2 years, a reasonable assumption given that a school class is involved, the mean would be increased.

12 The eighth score is 75.

13 The mean for the three groups is 57.1%.

14 The student must achieve 86% or more in the tenth item.

15 The mean of the other five scores is 50.

16 The girls achieved a mean of 62.7%.

17 The mean is the same as the median.

18 The mean is greater than the median.

19 The mean is less than the median.

20 The mean is the same as the median.

21 The mean is less than the median.

22 The mean is greater than the median.

23 The mean is the same as the median.

24 The mean is the same as the median.

25 The mean is greater than the median.

26 The mean is greater than the median.

27 The mean is less than the median.

28 The mean is the same as the median.

29 The mean is greater than the median.

30 The mean is less than the median.

ISBN 9780170390262

31 The mean is greater than the median.

32 The mean is greater than the median.

33 and **34** Answers not given here. Compare your answers with those of someone else in your class and discuss the merits of each.

Exercise 2B   PAGE 32

1 Mean 146, median 143, mode 137, range 40

2 Mean 82, median 83, mode 78, range 18

3 Mean 22.8, median 22, no mode, range 32

4 Mean 52.4, median 50, mode 35, range 45

5 Mean 2.3 (1 dp), median 2, mode 3, range 7

6 Mean 8.4 (1 dp), median 9, mode 6, range 7

7 Mean 18.8 (1 dp), median 19, mode 20, range 5

8 8.5   **9** 10.5   **10** 33.1   **11** 69.0

12 $447 600, $439 000. 8 are lower than the mean. 5 are lower than the median.

13 The mean number of bedrooms per house is 3.65.

14 a The modal salary is $68 000.

    b The median salary is $68 000

    c The mean salary is $73 640

15 The mean number of sunlight hours is 10.9 (1 dp) and the median is 11.1.

16 a The mean is 75.4 (1 dp).

    b The median is 77.

    c 10 students scored less than 60%.

    d 10% (3 of the 30) scored greater than 75%.

17 a There were 12 girls in the group.

    b The shortest girl was 148 cm tall.

    c The mean height for the girls is 159 cm (nearest cm).

    d The mean height for the boys is 167 cm.

    e The mean height for the 29 students is 164 cm (nearest cm).

    f

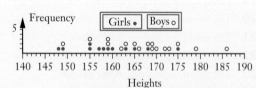

18

Score	21 - 25	26 - 30	31 - 35	36 - 40	41 - 45	46 - 50
Freq.	4	7	7	9	13	10

    a 37.8   **b** 38

19 a $40 \leq h < 50$ is the modal class

    b Mean = 44.7 hours

20 a Median lies in $230\,000 \leq C < 240\,000$

    b Mean price = $240 000

Exercise 2C   PAGE 37

(Answers are given to suggest typical comments that could be made.)

1 The displayed data involves 64 test marks altogether.

Using the mid-point of each class interval gives an estimated mean of 61.25, the median lies in the 60 to 70 interval and the modal class is the 70 to 80 interval.

The marks are spread from about 20 to about 100, a range of approximately 80, but almost all of the marks (61 out of the 64) are actually spread from about 40 to 80.

The 61 marks from 40 to 80 are reasonably evenly spread amongst the four intervals 40 to 50 (15 marks), 50 to 60 (13 marks), 60 to 70 (16 marks) and 70 to 80 (17 marks).

There were no marks between 30 and 40 and none between 80 and 90 so with 1 mark between 20 to 30 and the 2 marks between 90 and 100 the histogram features two 'gaps'.

2 The displayed data involves 82 lengths altogether.

With length involved it is likely that each class interval shows lengths rounded to the nearest whole centimetre. Using these whole centimetre values gives a mean length of 4.3 cm (to one decimal place).

The median length is in the 3.5 cm to 4.5 cm interval and this is also the modal class.

Whilst the lengths are spread from about 2.5 cm to about 9.5 cm, a range of approximately 7 cm, almost all (77 out of the 82, approx 94%) are actually between 2.5 cm and 5.5 cm.

The 77 lengths from 2.5 cm to 5.5 cm are reasonably evenly spread amongst the three intervals centred around 3 cm (24 lengths), 4 cm (27 lengths) and 5 cm (26 lengths).

The histogram features a 'gap' with no lengths between 5.5 cm and 7.5 cm.

5 of the recorded lengths are unusually long compared to the other 77 lengths.

3 The displayed data involves 85 lengths altogether.

With length involved it is likely that each class interval shows lengths rounded to the nearest whole centimetre. Using these whole centimetre values gives a mean length of 6.0 cm (to one decimal place).

The median length is in the 5.5 cm to 6.5 cm interval and this is also the modal class.

The lengths are spread from about 2.5 cm to 9.5 cm, a range of approximately 7 cm.

The distribution of scores is reasonably uniform with each 1 cm class interval containing roughly the same number of lengths (from a low of 10 to a high of 14).

The histogram has no gaps.

4 The displayed data involves 100 lengths altogether.

Using the mid-point of each class interval gives an estimated mean of 37.1 cm.

The median length lies in the 34.5 cm to 39.5 cm interval and this is also the modal class.

The distribution of the lengths is very symmetrical in nature with a tall central column and the frequencies falling away on either side of this centre.

Whilst the lengths are spread from a low of 19.5 cm to a high of 54.5 cm, a range of 35 cm, seventy three of the 100 lengths are between 29.5 cm and 44.5 cm, a range of just 15 cm, i.e. most lengths are close to the central mean and median values.

5 The tabulated data involves 96 scores altogether.

Using the centre of each class interval gives an estimated mean of 30.3 (to one decimal place). The median lies in the 26–30 interval and the modal class is the 21–25 interval.

Although the scores range from about 21 to about 55, a range of 34, the first interval (scores of 21–25) has the highest frequency with approximately 36% of the

scores, and the frequencies then decrease as the scores increase.

Almost 60% of the scores are from 20.5 to 30.5 and just less than 5% are from 45.5 to 55.5.

6 The displayed data involves 106 scores altogether.

Using the centre of each class interval gives an estimated mean for the scores of 34.7, to one decimal place. The median lies in the 30 to 40 interval.

The scores are spread from about 0 to about 70, i.e. a range of approximately 70.

The scores are reasonably symmetrically spread about a mid-point of about 35. Frequencies rise on either side of this centre to give a bi-modal appearance peaking at the first interval, 28 scores for which $0 \le$ score < 10, and again at the last interval, 25 scores for which $60 \le$ score < 70. Half of the scores are either in this first interval or in the last interval.

Miscellaneous exercise two   PAGE 39

1 **a** $6x + 15$ **b** $35x - 15$ **c** $-2 + 10x$
    d $22x - 9$ **e** $-2x + 11$ **f** $4x + 9$
    g $x + 5$ **h** $14x - 1$

2 Nominal categorical.

(With just two categories, Yes and No, it could also be referred to as binary, or binomial data.)

3

	Advantages	Disadvantages
Mean	Commonly understood as what we mean by the average. Output from statistical calculators. Every score in the set is used when determining the mean. May not be the central score but likely to be reasonably central if no outliers.	Can be greatly influenced by extreme scores (outliers). May not be one of the scores itself. (For example the mean number of children per married couple may not itself be a whole number of children.)
Median	As many scores below the median as above so it is central. Not affected by extreme scores. Output from statistical calculators. Easily calculated for small amounts of data. Can simplify measuring tasks - for example to find a median height we need only place the items in order and measure the middle one whereas for the mean we would need to measure all. To determine the median time for a team of 5 cyclists we can stop the clock when the third one crosses the line, we don't need to time them all.	May not be one of the scores itself. Can be tedious to rearrange a lot of scores in order if doing it manually.
Mode	It will be one of the scores. Not affected by an isolated outlier. Easy to work out for small data sets. Output from statistical calculators. Gives the most common score.	Not necessarily at all central. There may not be a mode.

4 The student needs at least 102% in test eight , i.e. cannot pass the course.

5 Answers not given here. Compare your answers with those of someone else in your class and discuss the merits of each.

6 Many possibilities, for example: $14 000, $15 000, $16 000, $17 000, $18 000, $88 000.

7 a Males 20 – 25: 76·7. Males 65 – 70: 85·7. Females 20 – 25: 69·5. Females 65 – 70: 83·5.

b

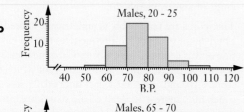

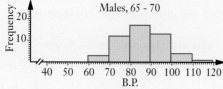

c

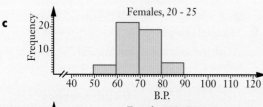

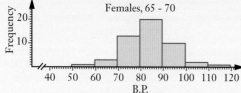

d For both sexes the 65 to 70 year olds generally have higher diastolic blood pressures than the 20 to 25 year olds suggesting diastolic blood pressure may increase with age. In both age groups the mean male diastolic blood pressure is higher than the mean for the corresponding female group.

Exercise 3A   PAGE 45

1 a 31   **b** 42   **c** 38 000
2 a 12.8   **b** 18.75
3 a 0   **b** 4.72 (2 d.p.)
   c 10.12 (2 d.p.)   **d** 7.93 (2 d.p.)
4 a B   **b** A
5 a B   **b** A
6 a Neither mean is greater, both = 5.
   b A

7 a A
   b Neither standard deviation is greater, they are both the same.
8 a II   **b** II   **c** II   **d** I
9 Range = 24, mean = 17, standard deviation = 6.58 (2 d.p.).

Exercise 3B   PAGE 49

Note: The answers below give σ_n when the standard deviation of a given set of scores is asked for. However it is recognised that in some states σ_{n-1} may be expected to be given whenever the standard deviation is requested. For this reason, if to the given accuracy the two values differ the σ_{n-1} value is shown in brackets.

1 Mean 12.5,   Standard deviation 1.7   (1.9)
2 Mean 25.4,   Standard deviation 13.5   (14.2)
3 Mean 31,   Standard deviation 1.7   (1.9)
4 Mean 6.99,   Standard deviation 0.47   (0.49)
5 Mean 30.2,   Standard deviation 3.7
6 a i C   **ii** B
    iii A   **iv** B
   b A: Mean = 8.1, Standard deviation = 1.7 (1.8)
      B: Mean = 3,   Standard deviation = 0.8 (1 dp)
      C: Mean = 5.5, Standard deviation = 2.9 (1 dp) (3.0)
7 a 5.76, 0.26   (0.28)   **b** 5.83, 0.07   (0.08)
8 a Yes   **b** Yes
9 a Mean = 160.44 cm,
     Standard deviation = 8.87 cm   (9.06 cm)
   b Now use σ_{n-1} because sample is used to predict standard deviation of population.
     Estimated standard deviation of population = 9.06 cm.
10 More likely to be a small type B. 18 mm is a little more than 1 standard deviation from the type B mean which is more likely than being 3 standard deviations from the type A mean.
11 a Mean = 97.45 cm,
     Standard deviation = 8.346 cm (3 dp)   (8·452 cm)
   b 80%   **c** 92.5%
   d 97.5%   **e** 8·45 cm
12 a Mean = 155.3°C,
     Standard deviation = 11.74°C (2 dp)   (12.37°C)
   b Mean = 159.1°C (1 dp)
     Standard deviation = 2.81°C (2 dp)   (2.98°C)
13 31.5

14 a Mean = 30.4°C (1 dp),
Range = 19.3°C,
Standard deviation = 4.9°C (1 dp). (5.0°C)

b The particular year involved has a mean maximum daily temperature for December that is approximately 3°C higher than long term mean.

c Mean = 15.8°C (1 dp),
Range = 14.4°C,
Standard deviation = 3.3°C (1 dp).

d The particular year involved has a mean minimum daily temperature for December that is approximately 0.5°C lower than long term mean.

15 a Mean = 61.4 (1 dp), Standard deviation = 16.3 (1 dp) (16.4)

b

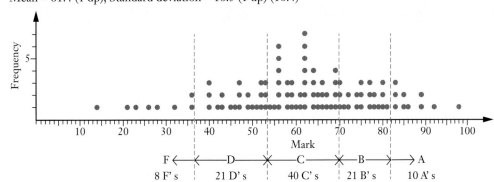

16 A passes. B rejected, condition 5. C passes. D rejected, condition 3.

Exercise 3C PAGE 54

Note: The answers below give σ_n when the standard deviation of a given set of scores is asked for. However it is recognised that in some states σ_{n-1} may be expected to be given whenever the standard deviation is requested. For this reason, if to the given accuracy the two values differ the σ_{n-1} value is shown in brackets.

1 Mean 3.1   Standard deviation 1.3
2 Mean 32.5   Standard deviation 8.1   (8.2)
3 Mean 32.5   Standard deviation 14.3   (14.5)
4 Mean 5.1   Standard deviation 1.8   (1.9)
5 Mean 70.8   Standard deviation 21.1   (21.2)
6 Mean 37.25   Standard deviation 8.0   (8.1)
7 Mean 18.6   Standard deviation 12.9   (13.1)
8 Mean 37.1   Standard deviation 12.9
9 Mean 62   Standard deviation 21.0   (21.4)
10 Mean 69.6   Standard deviation 16.0   (16.3)
11 a 1.8 standard deviations from mean
b 3.6 standard deviations from mean
12 Uncoated seeds:
Mean number of successes per tray = 34.8
Standard deviation = 6.6 (correct to 1 decimal place).

Coated seeds:
Mean number of successes per tray = 42.05
Standard deviation = 5.4 (correct to 1 decimal place).

The trays with coated seeds tend to produce more successes per tray with less variability in the number of successes.

13 With outlier: Mean 16.3 minutes
Standard deviation 10.9 minutes (11.1 minutes)

Without outlier: Mean 15 minutes
Standard deviation 8.3 minutes (8.5 minutes)

14 a Mean number of students per school 253, standard deviation 116 (117) (nearest integers).

b Mean number of students per school 248, standard deviation 107 (nearest integers).

Miscellaneous exercise three PAGE 58

1 0.012, 0.021, 0.1, 0.102, 0.12, 0.2, 0.201, 0.21.

2 $\dfrac{1}{100}, \dfrac{1}{5}, \dfrac{1}{3}, \dfrac{1}{2}, \dfrac{3}{5}, \dfrac{2}{3}, \dfrac{7}{10}, \dfrac{3}{4}$.

3 a 11   **b** 26   **c** 28
d 4   **e** 16   **f** 40

4 A (2, 3),   B (5, 4),   C (3, 0),   D (0, 5),   E (2, –3),
F (5, –4),   G (–2, 3),   H (–5, 2),   I (–4, –2),   J (–2, –4).

5 **a** Nominal categorical.

  b For the year in question there were approximately 778 000 apple trees in W.A.

  c 2507 tonnes of peaches were produced commercially.

  d **i** The gross value per tonne for oranges was $345

    ii The gross value per tonne for nectarines was $1000

  e On average each tree yielded 48 kilograms of apples.

6 Advantage: Easily determined.

  Disadvantage: Only two scores involved so variation amongst other scores not taken into account.

7 Mean 6.68, median 7, mode 9, standard deviation 2.32 (2.38)

8 The new mean is 83.6.

9 The mean birth weight is 3.00 kg (to 2 dp)

10 **a** The survey involved people who were trading in an old vehicle and so all those surveyed had owned at least one vehicle.

  b For those surveyed the mean number of vehicles owned prior to the latest purchase is indeed 4.4, to one decimal place, but the two outlying values significantly influence this value. Quoting the median value of 3 vehicles owned prior to the latest purchase may be more representative. Also the median has the advantage of being an integer value and it may be preferable to quote the average as a whole numbers of cars.

11 Each short wire is to be made of length 36.6 metres

  Each medium wire is to be made of length 67.6 metres.

  Each long wire is to be made of length 85.4 metres.

12 **a** Mean 43.92, Standard deviation 13.99 (14·28)

  b Unit C: 1.5,   Unit E: 0.7,

    Unit A: 0.649 (0·636), Unit B: –0.5,

    Unit D: –0.6

Exercise 4A PAGE 66

1 a 26   **b** 15   **c** 30   **d** 4

  e 36   **f** 15

2 a 20   **b** 12   **c** 26   **d** 6

  e 36   **f** 14

3 a 14   **b** 12   **c** 16.5   **d** 8

  e 19   **f** 4.5

4 a 61   **b** 50   **c** 80   **d** 20

  e 93   **f** 30

5 a C   **b** D   **c** A   **d** C

  e B   **f** C   **g** B

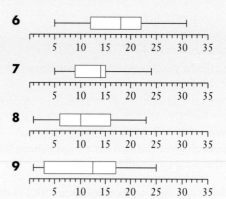

10 **a** The statement is incorrect. There will be the same number of scores below the median as there are above it.

  b If we rank spread on the basis of range (32 for class I and 31 for class II) or interquartile range (14 for class I and 13 for class II) then we could say that the class I marks are more spread out than the class II marks. However, even if we only consider these measures there is not much in it. Furthermore box plots do not show us all of the individual marks and the spread summaries of range and interquartile range are each determined using only two scores. Using other criteria to judge spread, for example standard deviation which uses all of the scores in its determination, we could well end up disagreeing with the given statement.

  c Considering only the range of scores, 34 for class III and 32 for class I, might tempt us to suggest that class III marks are more spread out but the interquartile range, 9 for class III and 14 for class I, suggests the opposite conclusion. The long lower whisker in the class III boxplot could be caused by just one outlying score. Thus whilst the statement could conceivably be correct under some suitably chosen definition of spread it would be unwise to claim the statement true under a more general understanding of 'more spread out'.

  d Based on the 'five point' nature of the information a box plot provides (i.e. lowest, Q1, median, Q3, highest) the statement seems correct and justified. Were we to know the individual scores we might find the distribution of scores within each quarter differs markedly between the two classes but without this information the statement is a reasonable statement to make based on the boxplot data.

  e Based on this test the top student in class III would be about 25% of the way down the rank positions of the students in class I, not necessarily the 25th student. Hence the statement is not one that can be concluded from the given information. Box plots do not tell us the number of data points involved.

f Class II certainly had at least one student who scored a lower mark than the lowest mark from the other two classes but we cannot conclude that there were 'lots' of students for whom this could be said.

Exercise 4B PAGE 72

(Answers are given to most questions to suggest some typical comments that could be made. For the questions for which comments are not given compare and discuss your descriptions with those of others in your class.)

Note: The answers below give σ_n when the standard deviation of a given set of scores is asked for. However it is recognised that in some states σ_{n-1} may be expected to be given whenever the standard deviation is requested. For this reason, if to the given accuracy the two values differ the σ_{n-1} value is shown in brackets.

1 The displayed data involves 42 test marks altogether.

The distribution of marks give a mean mark of 70, the median lies in the 70 to 80 interval and the modal class is also the 70 to 80 interval.

The marks are spread from about 10 to about 90, a range of approximately 80, and the standard deviation is 16.4. (16.5)

The distribution of marks is skewed to the left with 26 of the 42 marks (almost 62%) being 70 or more and 35 of the 42 (approximately 83%) being 60 or more. The distribution features a 'gap' with no marks between 30 and 50 but 3 marks were between 10 and 30.

2 The displayed data involves 100 lengths altogether.

The mean length is 5.97 cm and the median length (nearest centimetre) is 6 cm.

Lengths are spread from a low of 3 cm to a high of 9 cm, a range of 6 cm. (However the measurements probably involve rounding so the low could be 2.5 cm and the high could be 9.5 cm to give a range of 7 cm.) The standard deviation of the lengths is 2.3 cm.

The distribution of the lengths is approximately symmetrical in nature about the central value of 6 cm. It is bimodal with peaks at 3 cm and 9 cm. Seventy eight of the 100 recorded lengths are either between 2.5 cm and 4.5 cm or between 7.5 cm and 9.5 cm. The remaining 22 lengths fall in the 4.5 cm to 7.5 cm interval. Hence most of the lengths are situated away from the central mean and median values.

3 The displayed data involves 104 scores altogether.

The distribution of scores give a mean score of 42.0, the median lies in the 30 to 40 interval and the modal class is also the 30 to 40 interval.

The scores are spread from about 10 to about 100, a range of approximately 90, and the standard deviation is 18.2. (18.3)

The distribution of scores is skewed to the right with 74 of the 104 scores being between 10 and 50 and the remaining 30 being between 50 and 100. Just 9 of the 104 scores were between 70 and 100.

4 Comments not given here.

Compare and discuss your descriptions with those of others in your class.

5 The tabulated data involves 100 scores altogether.

The distribution of scores give a mean of 48.05, the median lies in the 46–50 interval and the modal class is the 41–45 class.

The scores are spread from about 31 to about 65, a range of approximately 34, and the standard deviation is 10.04. (10.09)

The distribution of scores is uniform with each class interval containing roughly the same number of scores (from a low of 13 to a high of 16).

6 The tabulated data involves 100 scores altogether.

The distribution gives a mean score of 15.1, the median lies in the $10 \leq x < 20$ interval and is probably much nearer to 10 than it is to 20. The modal class is the $0 \leq x < 10$ interval.

The scores are spread from about 0 to about 70, a range of approximately 70, and the standard deviation is 12.9. (13·0)

With almost half of the scores in the $0 \leq x < 10$ interval and the frequencies decreasing as we move right, the distribution of scores is skewed to the right. Eighty eight of the 100 scores are such that $0 \leq$ score < 30 whilst just 6 are such that $40 \leq$ score < 70.

7 The median of data set A (27) is higher than the median of data set B (24).

With the lower quartile at 20 and the upper quartile at 30 in each set the middle 50% of the data points in each set are spread over the same scores and each data set has an interquartile range of 10. The range of the two data sets is very similar, 34 for set A and 35 for set B. However for data set A the left whisker is the longer whisker, and there is more of the box to the left of the median than the right, indicating that data set A is skewed to the left. Conversely data set B has the right whisker noticeably longer than the left, and more of the box to the right of the median indicating that data set B is skewed to the right.

8 Comments not given here.

Compare and discuss your comments with those of others in your class.

9 Assuming the location of the Regional Meteorology Station to be typical for the region as a whole the region experienced rain on 110 days of the year, i.e. approximately 30% of the days in the year had some rain.

On more than half of the rainy days the rainfall was less than 5 mm. The total rainfall for the year was approximately 775 mm with approximately 8% of this total falling on just one day. If we include the days on which no rain fell the average daily rainfall for the year was approximately 2.1 mm per day. Considering only days on which rain fell the average daily rainfall was approximately 7 mm per rainy day. If we discount the one day of unusually high rainfall these averages become 1.95 mm and 6.5 respectively. For the days that rain fell the rainfall figures have a standard deviation of 8.1 mm but if we discount the one day of unusually high rainfall this falls to 6.1 mm.

The overall distribution is skewed to the right with one extreme value in the 60 to 65 mm interval and all other rainy days recording less than 30 mm.

10 The distribution of scores of the 196 students give a mean score of 83.8. The median score is in the 81 to 90 interval and the modal class is the 91 to 100 interval with approximately 21% of the scores in this interval.

151 of the 196 students (77% of them) achieved a score of over 70 in the exam (remember though that this 70 is a raw score out of at least 120 and is not a percentage score) and 119 of them achieved a raw score over 80.

The students achieved scores from a low of about 21 to a high of about 120, hence the range of scores was approximately 99. The scores had a standard deviation of 20.4.

With the distribution showing a long tail to the left the scores were skewed to the left.

11 The 76 donors involved had a mean age of 36 (nearest year), standard deviation 14.

Their ages ranged from around 15 to almost 60 with a median age of approximately 40.

The 76 recipients involved had a mean age of 44 (nearest year), standard deviation 12.

Their ages ranged from around 15 to mid sixties with a median age of approximately 45.

Thus whilst the ages of the donors and the recipients were spread across similar age ranges, with just six of the recipients older than the oldest of the donors, the donors tended on average to be younger than the recipients.

The age distribution of the donors is bimodal, peaking around 20 and again around 50, and is roughly symmetrical rising on either side of a low frequency central age of approximately 35. In contrast the age

distribution of recipients is negatively skewed with frequencies tailing off to the left of the modal age of about 50. Over half of the recipients were between 42 and 58.

Approx 40% of donors but only 15% of the recipients (approx) were aged under 30.

Miscellaneous exercise four   PAGE 74

1 With the outlier: Mean = 10
Standard deviation = 8.67 (9.19).

Without the outlier: Mean = 7
Standard deviation = 1.87 (2).

2 a 1   **b** 1   **c** 2   **d** 3
e 1   **f** 2   **g** 75%   **h** 2

3 The standard deviation will increase.

If we remove scores that are close to the mean, as those in the central column are, the average distance from the mean will increase. Hence the standard deviation will increase.

4 Ask others in your class to read and comment on your article and you do the same for theirs.

5 A with boxplot 4.   B with boxplot 1.
C with boxplot 2.   D with boxplot 3.

6 a Any scores less than 10 or greater than 58.

b Any scores less than 3 or greater than 75.

c The scores of 10, 17 and 70 are outliers under the given definition.

d The scores of 42, 49 and 50 are outliers under the given definition.

7

Mark	0	1	2	3	4	5	6	7	8	9	10
Frequency	3	0	5	7	13	14	15	21	18	17	7
Cumulative frequency	3	3	8	15	28	42	57	78	96	113	120

a 17   **b** 113
c 35%   **d** One eighth
e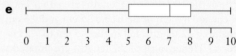

f

(histogram of Frequency vs Mark)

g The marks have a mean of 6.31 and a median of 7. The modal score is also 7.

With the lowest score of zero (3 students) and highest of 10 (7 students) the range of the scores is 10. No student scored a total of 1 but all other possible totals featured.

With a lower quartile of 5 and an upper quartile of 8 the interquartile range is 3.

The standard deviation is 2.35 (2·36) and the distribution is skewed to the left as can be seen from the histogram, as the longer box and whisker to the left of the median than the right suggests, and the fact that mean (6.31) is less than median (7) further reinforces.

8 3.85 (4.30)

Miscellaneous exercise five   PAGE 81

1 Histogram with new class intervals not shown here.

Histogram shows that the apparent uniform distribution of times is not the case. The new histogram indicates a distribution that dips on either side of two peaks - i.e. a more bi-modal shape to the distribution. Distribution is reasonably symmetrical about the time of 50 seconds. The apparent uniformity of the original distribution misses the bi-modal feature because the small number of class intervals cause some detail to be lost.

2 a $7x + 23$ **b** $16x + 17$ **c** $7x + 2$
  d $x - 10$ **e** $5x + 18$ **f** $x + 2$

3 a 15 **b** 44 **c** 82
  d 25 **e** 44.7 **f** 15.6 (15·7)

4 a Some of the newly diagnosed sufferers will be counted in more than one category, for example a male smoker under 30.

  b No. Statement not necessarily true. The graphs show *percentage* of newly diagnosed sufferers in each category, the overall *number* of male sufferers could have increased.

Exercise 6A   PAGE 93

1 a $x = 6$ **b** $x = -26$ **c** $x = 28$
  d $x = 6$ **e** $x = 4.5$ **f** $x = 3$
  g $x = 4$ **h** $x = 8$ **i** $x = 1$
  j $x = 13$ **k** $x = 5$ **l** $x = 1.4$
  m $x = 15$ **n** $x = 42$ **o** $x = 5$
  p $x = 1.4$ **q** $x = 84$ **r** $x = 7$
  s $x = 2$ **t** $x = -1$ **u** $x = 12$
  v $x = 2$ **w** $x = -0.5$ **x** $x = 18$
  y $x = 13$

2 a $P = 650$ **b** $A = 1335$ **c** $I = 55$
3 a $v = 20$ **b** $u = 18$ **c** $a = 2$
  d $a = -3$ **e** $t = 3$ **f** $t = 11$
4 a $r = 3.98$ **b** $r = 15.12$ **c** $C = 50.27$
  d $r = 64$
5 a $A = 25.13$ **b** $r = 2.84$ **c** $h = 5.31$
6 a $t = 14$ **b** $u = 9$ **c** $v = 15$
7 a $R = 11$ **b** $R = 45$
  c $R_1 = 10$ **d** $R_3 = 27$
8 a $h = 5.95$ **b** $h = 35.79$
9 a $V = 25$ **b** $P = 10$
  c $V = 20$ **d** $P = 2.5$
10 a $P = 735$ **b** $P = 7840$
  c $m = 150$ **d** $h = 2.25$
11 a The cost is $570.
  b The cost is $360.
  c The person could travel 2500 km, or less.
12 a The commissions received are $3100, $3400 and $2100 respectively.
  b The house was sold for $445 000.
  c The house was sold for $850 000.
13 a The male has an estimated height of 1.7 m.
  b The expected humerus length is 40 cm.
14 a The profit will be $2000.
  b The profit will be $14 700.
  c The profit will be $23 590.
  d The least number is 3052.
  e The firm would make a loss of $6890.
  f The greatest profit is $33 750.
  g The firm would lose $29 750.
  h The firm must sell at least 2343 calendars to avoid making a loss.

Miscellaneous exercise six   PAGE 96

1 a $x = 7$ **b** $x = 2$ **c** $x = 3.5$
  d $x = 2.5$ **e** $x = 14$ **f** $x = 7.5$
2 a $A = 69$ **b** $h = 2.5$
3 $a = 3, b = 8, c = 10, d = 12, e = 14$.
4 The student requires a mark of 47% or more in test six.
5 a A **b** C **c** B **d** A
  e D and B (in that order)
6 5 A, 9 B, 21 C, 10 D, 2 Fail.

7 The median would be the fairest to use as it will not be unduly influenced by the occasional sale of one of the luxury properties. It might suit the agent to be able to quote a higher average by using the mean, with one or more luxury properties included in the calculation, but for a steadier monthly value that allows a general trend to be observed, free from the occasional monthly 'spike' when a luxury property is sold, the median would be more suitable. The mode might occasionally be central but not necessarily, so that too would not be as good as the median.

8 a Mean $65 588   Standard deviation $16 488   ($16 532)

  b Mean $66 098   Standard deviation $16 393   ($16 441)

9 Six applicants are invited for interview.

10 Discuss your descriptions with those of others in your class and with your teacher.

11 a Question cannot be answered from the given information. The graphs indicate percentage of the population in the various age intervals, not population numbers.

  b Approximately nine million of country A's population are aged 70 or over.

  c Let others read and comment on your report and you read and comment on their report.

Exercise 7A   PAGE 104

1 11   **2** 12   **3** 2.3   **4** $1\frac{2}{3}$

5 9   **6** 2.75

7 a 1   **b** 5   **c** 3

  d add starting number   **e** divide by two

  f

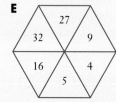 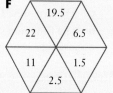

   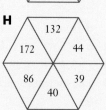

 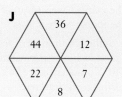

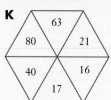 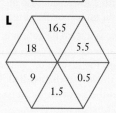

Exercise 7B   PAGE 108

1 a $5x + 6$ **b** $14 - x$ **c** $5(x + 6)$

  d $2x - 7$ **e** $2(x - 7)$ **f** $3(2x + 5)$

2 A: $2x + 1 = 10, x = 4.5$ B: $2(x - 1) = 10, x = 6$

  C: $10 - x = 1, x = 9$ D: $2(x + 1) = 10, x = 4$

  E: $\frac{x}{2} - 1 = 10, x = 22$ F: $x - 10 = 1, x = 11$

  G: $\frac{x - 1}{2} = 10, x = 21$ H: $2x - 10 = 1, x = 5.5$

3 The number first thought of was 9.

4 The number first thought of was 3.

5 The number first thought of was 10.

6 The number first thought of was 5.

7 The number first thought of was 9.

8 The number first thought of was 5.

9 The number first thought of was 7.

10 The number first thought of was 13.

11 The number first thought of was 9.5.

12 The number first thought of was 12.

Exercise 7C   PAGE 111

1 Bob contributes $6500, Tony contributes $12 000.

2 Sue should receive $12 000, Lyn $18 000 and Paul $17 000.

3 Bill is 43 years old now.

4 a $(5000 + 22x)$ **b** 278

5 a $3x$ hours

  b $4(120 - x)$ hours

  c 75 standard and 45 deluxe.

6 a The width should be 85 m.

  b The length should be 95 m.

  c The area should be 8075 m^2.

ISBN 9780170390262

7 $(5x - 2)$ years. Heidi is 9 years old now.

8 They need 240 tickets at \$12 each and 610 at \$8 each.

9 a i $x + 2000$

    ii $2(x + 2000)$

  b 3000 acres of lupins, 5000 acres of barley and 10 000 acres of wheat.

10 a i $150x$ grams

    ii $80(50 - x)$ grams

  b Each 50 kg of *QuickGrow* should contain 32 kg of X and 18 kg of Y.

11 The original order was for 185 hardback and 115 softback.

12 \$2800 was invested with company A and \$2200 was invested with company B.

Exercise 7D   PAGE 116

1 a $a = 6$ **b** $b = 4$ **c** $c = 6$

  d $d = 7.5$ **e** $e = 3.6$ **f** $f = 2.8$

  g $g = 1.5$ **h** $h = 1.7$ **i** $i = 4.8$

2 \$4200 needs to be invested.

3 7.5%

4 The initial investment was \$850.

5 Four and a half years.

6 146 days

7 Annual rate of 6.5% required.

8 $R = 8.2$

9 6% is the required annual rate of simple interest.

10 There were 1463 females in the audience.

11 There are 391 female students in the school.

12 The height of the tree is approximately 13.5 metres.

Note: Using ratios an 'exact' answer of 13.5 metres is obtained. However the word approximately is used in the answer because the nature of the situation means that this answer would have to be regarded as being something of an approximation. Will the highest point of the tree be directly above the centre of the base? Will the end of the shadow be somewhat blurry and difficult to locate accurately? (This applies to some later questions in this exercise too.)

13 The building is approximately 25 metres tall.

14 $h = 3.5x$. If $x = 1.5$, $h = 5.25$.

15 The flagpole is approximately 7 metres tall.

16 The river is approximately 54 metres wide.

17 The pylon is approximately 8.6 metres tall.

18 Mary borrowed \$1200 in the first place.

19 Mai borrowed \$6800 in the first place.

Miscellaneous exercise seven   PAGE 121

1 There are 342 female students in the school.

2 a $x = 4.2$ **b** $x = 7.5$

  c $x = 8.75$ **d** $x = 2.8$

3 $m = 15$

4 8

5 a 86   **b** 79   **c** 7   **d** 9

  e Naomi is correct in her statement that the range of the male scores is bigger than the range of the females. ($Range_{Male} = 9$, $Range_{Female} = 7$).

However the range depends only on the lowest score and the highest score so it is unwise to claim that the male scores as a whole are more spread out than the female scores based solely on this fact. Indeed other measures of spread namely the interquartile range (5 for female and 2 for male) and the standard deviation (2.4 for female and 2.0 for male) disagree with Naomi's statement and suggest that the female scores are more spread out than the male scores.

6 The number first thought of was 18.

7 At least \$14 650 needs to be invested for the account to be at least \$19 000 in 5 years.

8 \$54 000 into the account paying 6.3% and \$26 000 into the account paying 5.4%.

9 a 70.5

  b Mean 68   Standard deviation 26.5

10

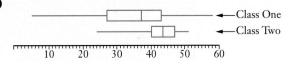

Overall class two performed better even though top mark was in class one:

The median for class one (37) is lower than the median for class two (43.5).

Having the individual scores available we can also determine the means: 34.7 for class one and 42.5 for class two.

The upper quartile in class one is lower than the median in class two.

The marks of class one are more spread out than those of class two:

The range of class one (53) greatly exceeds the range of class two (27) as does the interquartile range (16 for class one compared to 7 for class two).

Having the individual scores available we can also determine the standard deviations: 12.2 (12.5) for class one and 6.2 (6.3) for class two.

(Answer continued over page.)

The box plot for class one suggests a reasonably symmetrical distribution perhaps skewed left a little as suggested by the longer left whisker and the greater part of the box being to the left of the median. The class two boxplot also suggests a skew to the left because of the longer left whisker though within the box the median is centrally placed. (Such skewness is further suggested by the fact that mean < median in each class.)

The two distributions were based on a very similar number of data points, 25 for class one and 22 for class two.

11 Many possible answers but all must have:
- a total of 50 scores,
- at least one score in the $10 \to 20$ interval (as lowest score was 16),
- the 13th score (counting from the low end) in the $20 \to 30$ interval (as 1st quartile = 26),
- the mean of the 25th and 26th score in the $40 \to 50$ interval (as median was 46),
- the 13th score (counting from the top end) in the $50 \to 60$ interval (as 3rd quartile = 56),
- at least one score in the $70 \to 80$ interval (as highest score was 72).

One possibility is shown below:

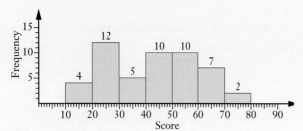

12 $x = 64\,000$.

Exercise 8A   PAGE 129

1	2	**2**	0.5	**3**	1	**4**	–1
5	–0.5	**6**	2	**7**	–0.5	**8**	3
9	–2	**10**	1	**11**	2	**12**	–1
13	–2	**14**	$-\frac{1}{3}$	**15**	5	**16**	1
17	3	**18**	0.2	**19**	5	**20**	–20

21 Points would lie in a straight line. Gradient = 2.
22 Points would lie in a straight line. Gradient = –2.
23 Points would not lie in a straight line.
24 Points would not lie in a straight line.
25 Points would lie in a straight line. Gradient = 5.
26 Points would not lie in a straight line.
27 Points would lie in a straight line. Gradient = –5.
28 Points would not lie in a straight line.
29 Points would lie in a straight line. Gradient = –3.
30 Points would lie in a straight line. Gradient = 3.
31 Points would lie in a straight line. Gradient = 2.
32 Points would lie in a straight line. Gradient = 3.
33 Points would not lie in a straight line.
34 Points would lie in a straight line. Gradient = 2.5.

Exercise 8B   PAGE 137

	a		**b**		**c**	
1	1		(0, 2)		$y = x + 2$	
2	2		(0, –3)		$y = 2x - 3$	
3	–1		(0, 3)		$y = -x + 3$	
4	0.5		(0, 3)		$y = 0.5x + 3$	
5	–3		(0, 2)		$y = -3x + 2$	
6	–2		(0, –3)		$y = -2x - 3$	
7	40		(0, 10)		$y = 40x + 10$	
8	5		(0, 4)		$y = 5x + 4$	
9	6		(0, 15)		$y = 6x + 15$	
10	–10		(0, 70)		$y = -10x + 70$	

11 A: $y = 7$,   B: $y = 4$,   C: $y = 1$,   D: $y = -4$, E: $x = -4$,   F: $x = -3$,   G: $x = 2$,   H: $x = 4$.
12 Relationship is linear. $y = 3x + 1$
13 Relationship is linear. $y = -4x + 25$
14 Relationship is linear. $y = 5x - 3$
15 The relationship is not linear.
16 Relationship is linear. $y = 1.5x + 0.5$
17 The relationship is not linear.
18 Relationship is linear. $y = x + 2$
19 The relationship is not linear.
20 Relationship is linear. $y = -x + 13$
21 Relationship is linear. $y = 2x + 21$
22 Relationship is linear. $y = 3x + 4$
23 Relationship is linear. $y = 5x - 12$
24 Relationship is linear. $y = 2x - 2$
25 Relationship is linear. $y = 3x + 1$
26 **a** 5   **b** (0, –10)   **c** $y = 5x - 10$
27 **a** –6.25   **b** (0, 37.5)   **c** $y = -6.25x + 37.5$

28 Table:

x	−4	−3	−2	−1	0	1	2	3	4
y	−5	−2	1	4	7	10	13	16	19

Rule: $y = 3x + 7$

Graph:

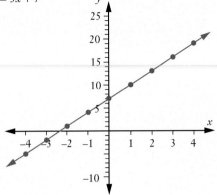

29 Table:

x	−4	−3	−2	−1	0	1	2	3	4
y	22	18	14	10	6	2	−2	−6	−10

Rule: $y = -4x + 6$

Graph:

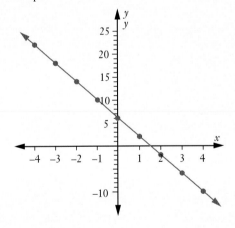

30 Table:

t	−4	−3	−2	−1	0	1	2	3	4
r	−2	0	2	4	6	8	10	12	14

Rule: $r = 2t + 6$

Graph:

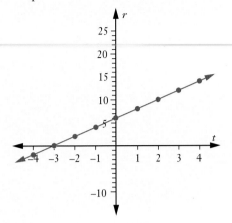

31 Table:

u	−4	−3	−2	−1	0	1	2	3	4
w	−9	−6	−3	0	3	6	9	12	15

Rule: $w = 3u + 3$

Graph:

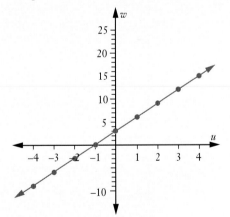

32 Table:

t	–4	–3	–2	–1	0	1	2	3	4
K	–7	–3	1	5	9	13	17	21	25

Rule: $K = 4t + 9$

Graph:

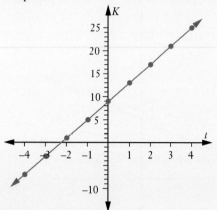

33 Table:

n	–4	–3	–2	–1	0	1	2	3	4
P	15	13	11	9	7	5	3	1	–1

Rule: $P = -2n + 7$

Graph:

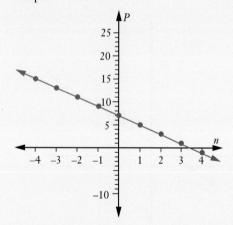

34

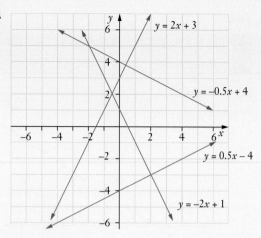

35 A: $y = -\dfrac{2}{3}x + 2$ B: $y = -1.5x + 3$

C: $y = 2.5x + 4$ D: $y = 4x$

E: $y = 3x$ F: $y = 3x - 2$

36 A: $y = -2$ B: $y = 0.5x + 2$

C: $y = -x + 2$ D: $y = 2x - 2$

E: $y = 0.5x - 2$

37 A: $y = 10x + 60$ B: $y = 30$

C: $y = -10x + 30$ D: $y = 30x$

E: $y = 30x - 90$

38 The graphs appear different because the scales used on the axes are not the same in the two graphs. Hence the intercepts with the y-axes appear different but in both cases will be at $(0, 3)$ and the intercepts with the x-axes appear different but in both cases will be at $(-1.5, 0)$.

Exercise 8C   PAGE 145

1 a 2   **b** 4   **c** 2   **d** 0.5

   e 0.25   **f** 1   **g** –2   **h** 2.5

   i 0.5   **j** –4   **k** 2.5   **l** 0.5

2 a Gradient 3   Cuts y-axis at $(0, -17)$

   b Gradient –2   Cuts y-axis at $(0, 13)$

   c Gradient –7   Cuts y-axis at $(0, 5)$

   d Gradient $-\dfrac{2}{3}$ Cuts y-axis at $(0, 8)$

   e Gradient –0.4   Cuts y-axis at $(0, 1.6)$

   f Gradient $\dfrac{2}{3}$ Cuts y-axis at $(0, 3)$

   g Gradient –0.5   Cuts y-axis at $(0, 11)$

   h Gradient –2.5   Cuts y-axis at $(0, 15)$

   i Gradient –1.2   Cuts y-axis at $(0, 12)$

3 $y = 0$

4 $x = 0$

5 A does not, B does not, C does, D does not, E does.

6 H and I.

7 $y = 3x + 4$, Yes

8 $y = 0.5x + 2$, D and E

9 $a = 1, b = -1, c = -13, d = 0, e = 9, f = 0$.

10 a $y = x + 2$ **b** $y = -x + 5$

    c $y = -2x + 8$ **d** $y = 5x + 8$

    e $y = 0.5x + 5$ **f** $y = -0.5x - 1.5$

    g $y = 1.5x - 11.5$ **h** $y = -\frac{1}{3}x + \frac{4}{3}$

11 a $y = x + 3$ **b** $y = -4x - 1$

    c $y = -3x + 43$ **d** $y = 2x - 1$

    e $y = \frac{1}{3}x + \frac{5}{3}$ **f** $y = -2x + 4$

    g $y = \frac{5}{3}x + 4$ **h** $y = -5x + 5$

Exercise 8D   PAGE 147

1 a $F = 1.8C + 32$

    b The value of m tells us the increase in the Fahrenheit temperature for each 1 degree increase in the Celsius temperature.

    c $131°F$ **d** $14°F$ **e** $15°C$ **f** $-20°C$

    g Yes there is such a temperature. $-40°C = -40°F$.

2 a The value of k tells us the increase in the length of the spring for each kilogram suspended from it.

    b The value of L_0 tells us the length of the spring when no weight is suspended from it i.e. it is the unstretched length or natural length of the spring.

    c $k = 0.2, L_0 = 0.45$, 5 cm.

3 a A: $(-80, 20)$, B: $(120, 120)$, C: $(-100, 60)$, D: $(-60, -20)$, E: $(100, 160)$, F: $(140, 80)$.

    b $y = 0.5x + 60$

    c $y = -2x - 140$

    d $y = -2x + 360$

4 a The value of m tells us how much the amount to be paid increases in dollars for each extra unit used. It is the cost per metered unit.

    b The value of c tells us the cost in dollars that we are charged even if we use no units. It is the 'standing charge'.

    c $A = 0.24N + 40$ **d** $88

    e 175 units

5 a $y = 2.4x$ **b** $y = x$ **c** $y = 0$

    d $y = x - 7$ (or in real life units $y = x - 175$)

    e $y = -0.4x + 14$ (or $y = -0.4x + 350$)

6 The gradient of the line, m, gives the amount the cost of the job increases for each hour increase in the time taken to do the job. It is the hourly rate charged.

The vertical axis intercept, c, gives the 'call out fee' i.e. the amount that is charged for arriving at your door (the hourly rate then goes on top of this).

The rule is $C = 120T + 80$

7 $C = 4.90 + 1.85x$ **8** $V = 1000 - 0.2t$

9 $P = 75n - 800$ **10** $P = 8x$

11 $t = \dfrac{k}{3} + \dfrac{1}{3}$

12 $P = 4.5N - 3650$

    a The value of m is 4.5 and this tells us that each extra ticket sold raises the profit by $4.50.

    b The value of c is -3650 which tells us that if no tickets are sold the loss will be $3650.

    c $3100   **d** $6925   **e** 812

13 a 110, 540   **b** $1660

14 $N = 40t + 210$

The value of m is 40 which tells us that the membership is increasing at approximately 40 members per year.

The value of c is 210 which tells us that at the beginning of the 5 year period there were approximately 210 members in the club.

If the linear relationship continues then when $t = 10$ the membership would be approximately 610.

15 a $P = 15x - 3750$

    b The company needs to sell at least 917 copies for a profit of more than $10 000.

16 a The 5740 tells us that when the monitoring program started there were approximately 5740 of these animals thought to be in existence in the wild.

The 350 tells us that the numbers of these animals thought to be in existence in the wild is decreasing at approximately 350 per year.

    b Graph not shown here.

    c $(16.4, 0)$. If the rate of decline continues there will be none of these animals in existence in the wild approximately sixteen and a half years after the monitoring program commenced.

Miscellaneous exercise eight   PAGE 152

1 A: $x = 4$ B: $y = -3$

    C: $y = x$ D: $y = x + 2$

    E: $y = 2x + 4$ F: $y = -x$

    G: $y = 0.25x + 4$ H: $y = 0.5x + 1$

    I: $y = -0.5x - 1$

2 Equation 3 must be true.

3 a $a = 2.4$ **b** $b = 3.5$ **c** $c = 6$

   d $d = 3.8$ **e** $e = 13.5$ **f** $f = 13.5$

   g $g = 4.5$ **h** $h = 3\frac{5}{9}$

4 There are 196 females in this workforce.

5 The student needs at least 66% in the final unit.

6 Dot frequency 1 with Boxplot C

   Dot frequency 2 with Boxplot A

   Dot frequency 3 with Boxplot D

   Dot frequency 4 with Boxplot B

7 a 72 **b** 39 **c** 51

8

9 a The gradient of 7 means that for each degree rise in temperature the number of chirps per minute goes up by 7.

   In theory the −16 suggests that at 0°C the cricket would make −16 chirps! However 'negative chirps' is a rather meaningless concept. The equation is really only valid for $N \geq 0$.

  b For $N \geq 0$ we need the temperature to be greater than 2.3°C. Indeed, according to the rule, for at least one chirp per minute the temperature needs to be approximately 2.5°C.

  c i The cricket makes roughly 82 chirps per minute.

    ii The cricket makes roughly 180 chirps per minute.

  d i The temperature is approximately 31°C.

    ii The temperature is approximately 24°C.

Exercise 9A PAGE 159

1 For $x < 0$ $y = -x - 4$

   for $0 \leq x < 4$ $y = 2x - 4$

   for $x \geq 4$ $y = 4$

2 For $x \leq -4$ $y = 5$

   for $-4 < x < 0$ $y = 3$

   for $0 \leq x \leq 3$ $y = -x$

   for $x > 3$ $y = x - 3$

3

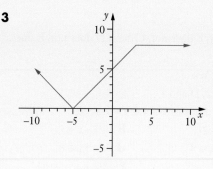

4

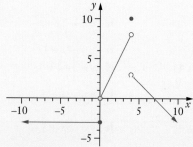

5 Ask someone in your class to read and constructively comment on your answer and you do the same for theirs.

6 a The blue broken line represents the journey of the cyclist.

  b The cyclist passes the walker between 8.52 a.m. and 8.53 a.m.

  c The walker took 30 minutes to walk to school.

  d The walker maintained a steady speed of 6 km/h.

  e The cyclist took 10 minutes to ride to school.

  f The cyclist maintained a steady speed of 18 km/h.

7 a The cyclist left town A at 7 a.m.

  b The cyclist reached town B at 11.20 a.m.

  c For the cyclist each stop was for 30 minutes.

  d i Prior to the first stop the cyclist maintained 20 km/h.

    ii Between the two stops the cyclist maintained 15km/h.

    iii After the second stop the cyclist maintained 30 km/h.

  e i From town A to town B the delivery truck maintained 60 km/h.

    ii From town B back to town A the delivery truck maintained 90 km/h.

  f When they were both travelling towards B the delivery truck passes the cyclist at about 9.35 a.m., and about 36 km from A.

  g When returning to A the truck passed the cyclist at about 11.05 a.m., and about 52 km from A.

8 (Graph not shown here.)

a The car reaches C at 11.54 a.m. and truck reaches town C at 12.15 p.m.

b From 8.30 a.m. to 9.30 a.m. the truck maintained a steady speed of 100 km/h.

c The average speed of the truck from A to B was 87 km/h (to the nearest km/h).

d The car passes the truck at 10.30 a.m. in town B, just as the truck is about to leave B.

9 a Someone with a taxable income of $30 000 would pay $7000 in tax.

b Someone with a taxable income of $40 000 would pay $11 000 in tax.

c Someone with a taxable income of $48 000 would pay $15 000 in tax.

d Someone with a taxable income of $3000 would pay no tax at all.

Someone paying tax of $20 000 would have a taxable income of $58 000.

10

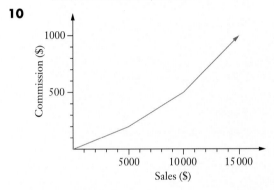

11 For the first $100 000 of the sale price the agent's fee is a fixed $6000.

From $100 000 to $300 000 the fee is $6000 plus 5% of the amount over $100 000.

From $300 000 and over the fee is $16 000 plus 2% of the amount over $300 000.

12 a

b

c

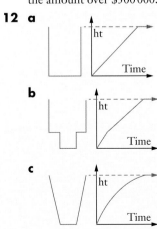

d

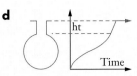

Miscellaneous exercise nine   PAGE 163

1 $P = 3t - 5$

2 $a = -1$, $b = 4$, $c = 9$, $d = 19$, $e = 29$, $f = 11$, $g = 99$

3 A: $y = -3$ B: $y = 1$
   C: $y = -0.5x + 5$ D: $x = 5$
   E: $y = x + 3$ F: $y = 9$
   G: $x = -3$ H: $y = 3x + 2$
   I: $x = 7$ J: $y = x$

4 a The 5200 and the 16 tell us that the fixed cost of production is $5200, even when no radios are produced, and then each radio produced adds $16 to this cost of production.

b When 100 radios are produced the mean cost per radio is $68.

c When 500 radios are produced the mean cost per radio is $26.40.

d When 1000 radios are produced the mean cost per radio is $21.20.

5 There are 210 year eight students in the school.

6 4

7 The number first thought of was 6.

8 There were 5 eights in the set.

9 a 33   **b** 37   **c** 16
   d 30 – 40
   Mean = 31.8

10 a 25%

b The range of the set B marks (39) exceed the range of the set A marks (32).

The interquartile range for set B (17) exceeds the interquartile range of set A (11).

c The median of those left in set B would be lower than that of set B before the move.

d The range of those left in set B would be less than that of set B before the move.

e The range of the set A scores would be unchanged.

f The interquartile range would be reduced.

Exercise 10A   PAGE 177

1 a 0.34   **b** 0.98   **c** 0.36   **d** 0.84
   e 3.08   **f** 0.60   **g** 0.77   **h** 0.77

3 a 11.5 **b** 66.4 **c** 52.4 **d** 17.5
e 75.5 **f** 53.1 **g** 63.4 **h** 25.8

4 a 1.3 **b** 3.4 **c** 2.9 **d** 20.5
e 14.0 **f** 12.6

5 a 23.6 **b** 44.4 **c** 54.5

6 a $\frac{a}{b}$ **b** $\frac{c}{b}$ **c** $\frac{a}{c}$ **d** $\frac{a}{b}$

e $\frac{c}{b}$ **f** $\frac{c}{a}$

7 $p^2 = q^2 + r^2$

8 a i 5.3 **ii** 5.27
b i 31 **ii** 31.0

9 2.6 **10** 11.2 **11** 9.8 **12** 9.0
13 4.1 **14** 6.3 **15** 16.5 **16** 12.8
17 9.6 **18** 28.7 **19** 65.8 **20** 3.5
21 51.1 **22** 56.5 **23** 6.5 **24** 10.8
25 9.2 **26** 3.7 **27** 40.8
28 $x = 16.8, y = 11.8$ **29** $x = 16.7, y = 33.7$
30 53.3

31 a 14.9 cm **b** 9.3 cm
32 a 36° **b** 8.6 cm
33 a The ladder reaches 7.06 m up the wall, to the nearest centimetre.
b The horizontal distance from the foot of the ladder to the wall is 3.76 m, to nearest cm.
34 a The ladder makes an angle of 66° with the ground, to the nearest degree.
b The light is 4.6 metres above ground, correct to one decimal place.
35 The kite is 41 m above ground level, to the nearest metre.
36 a AB makes an angle of 22° with the horizontal, to the nearest degree.
b AB is of length 2.15 m, to the nearest centimetre.
37 To the nearest centimetre, AB = 12.45 m, BF = 9.53 m and FC = 12.45 m.
38 θ = 38.7° correct to one decimal place.
The length of AD is 64% of the length of AE, to the nearest percent.
39 a Each short wire is of length 7.81 m (nearest cm) and makes an angle of 40° with the ground (nearest degree).
b Each long wire is of length 11.66 m (nearest cm) and makes an angle of 59° with the ground (nearest degree).
40 To the nearest centimetre, AC is of length 5.22 m, CG is of length 3.36 m, BH is of length 1.68 m, HC is of length 3.91 m.

41 To the nearest degree, each wire makes an angle of 62° with the horizontal.
42 a The largest possible length of the support wire EB is, to the nearest cm, 15.24 m.
b The shortest possible length of the support wire EB is, to the nearest cm, 12.56 m.
43 The bob rises 27 mm above its lowest position (to the nearest millimetre).
44 To the nearest metre point D is 10 metres above the horizontal ground.
45 The total length of steel required is 165 metres (to the nearest whole metre).
46 $x = 36.7, y = 6.1$

Exercise 10B PAGE 185

1 a 035° (or N35°E) **b** 080° (or N80°E)
c 110° (or S70°E) **d** 145° (or S35°E)
e 200° (or S20°W) **f** 300° (or N60°W)
g 215° (or S35°W) **h** 260° (or S80°W)
i 290° (or N70°W) **j** 325° (or N35°W)
k 020° (or N20°E) **l** 120° (or S60°E)
2 a 30° **b** 20°
c 35° **d** 15°
3 The height of the flagpole is 11.9 m, correct to one decimal place.
4 The angle of elevation of the sun is 26° (nearest degree).
5 The height of the flagpole is 10.0 m, correct to one decimal place.
6 Rounded up to the next metre the length is 19 metres.
7 To the nearest kilometre the ships are 10 km apart.
8 B is 66 m from A (to the nearest metre).
9 Ship C is approximately 26.9 km from ship A.
10 C is 216 m from A (to the nearest metre).
11 B is 91 m from A (to the nearest metre).
12 The height of the second tower is 65 metres, to the nearest metre.
13 The tree is approximately 14.8 metres tall.
14 The smoke is approximately 11.4 km from the first observation tower.
15 The ships are approximately 270 metres apart.
16 To the nearest metre the flagpole is 17 metres long.
17 The height of the flagpole is 15.5 metres, correct to one decimal place.
18 The required angle of elevation is 32° (nearest degree).

ISBN 9780170390262

Miscellaneous exercise ten PAGE 188

1 a $x = 9.4$ **b** $x = 18.7$ **c** $x = 56.3$
 d $x = 4.1$ **e** $x = 12.1$ **f** $x = 66.4$

2 The mean of the six amounts is $26 350 and the median is $14 100.

3 The mean for the girls was 21.2.

4 a $x = 21$ **b** $x = 2.5$ **c** $x = 7$
 d $x = 0.625$

5 a $s = 3$ **b** $a = 4$

6 The number first thought of was 15.

7 John borrowed $8600.

8 a

24	12
29	5

b

29	14.5
36.5	7.5

c

184	92
269	85

d

34	17
44	10

e

36	18
47	11

f

14	7
14	0

g

44	22
59	15

h

60	30
83	23

9 a Approximately 27%.
 b 37.9% of the income of the country is earned by the richest 10% of the population.

10 $y = 0.5x + 2.5$, $f = 7$, $g = -2$, $h = 13$, $i = -2$, $j = 4.4$.

11 a 59%   **b** More girls than boys.

12 A: $y = 5$ B: $x = -8$
 C: $y = 2x - 4$ D: $y = -x + 5$
 E: $y = 0.2x$ F: $y = x - 9$
 G: $y = 2x + 15$ H: $y = x - 12$
 I: $y = 2x - 13$ J: $y = -\frac{1}{3}x - 8$

13 a 49 500   **b** 22 200
 c 7.14 tonnes (to 2 d.p.)

14 $d = 11.44$ m, $c = 2.86$ m

Exercise 11A PAGE 193

1 5.4, 27.2 cm^2 **2** 45 cm^2
3 27 cm^2 **4** 20 cm^2
5 9.6 cm^2 **6** 211 m^2
7 8.8 cm^2 **8** 15.6 cm^2
9 18.1 cm^2 **10** 1730 mm^2

Exercise 11B PAGE 198

1 13.9 cm^2 **2** 75.8 cm^2 **3** 18.8 cm^2
4 101.8 cm^2 **5** 87.3 cm^2 **6** 3.8 cm^2
7 32.5 cm^2 **8** 23.0 cm^2 **9** 11.8 cm^2
10 17.3 cm^2 **11** 10.4 cm^2 **12** 6.8
13 7.3   **14** 6.6   **15** 67°
16 52°   **17** 32°
18 a $125 000   **b** $51 000

19 The second block has the greater area, by 9 m^2, to the nearest square metre.

20 The owner of block A receives $472 193.55 and the owner of block B receives $777 806.45.

21 AC needs to be 180 metres long, rounded up to the next whole metre.

Exercise 11C PAGE 207

1 5.8   **2** 6.6   **3** 12.9
4 54.8   **5** 46.0   **6** 54.2
7 58 (nearest integer)   **8** 12.3 (1 d.p.)
9 54 (nearest integer)   **10** 14 (nearest integer)
11 105 (nearest integer) **12** 126 (nearest integer)
13 6.7 (1 d.p.)   **14** 75 (nearest integer)

15 The pole is of length 614 cm, to the nearest centimetre.

16 The two shot journey is 38 metres further than the direct route, to the nearest metre.

17 7.1   **18** 1.7
19 21.8   **20** 73.0
21 59 (nearest integer)   **22** 14.4 (1 d.p.)
23 43 (nearest integer)   **24** 111 (nearest integer)
25 44 (nearest integer)   **26** 62 (nearest integer)
27 11.9 (1 d.p.)   **28** 146 (nearest integer)

29 The boat is then 13.4 km from its initial position, correct to one decimal place.

30 After eight seconds Jim and Toni are 10.7 metres apart, correct to one decimal place.

31 B is 13.1 km from C, correct to one decimal place.

32 75 (nearest integer)   **33** 99 (nearest integer)
34 617 (nearest integer) **35** 5.39 (2 d.p.)
36 135 (nearest integer) **37** 80 (nearest integer)
38 160 (nearest integer) **39** 54 (nearest integer)

40 The lengths of AC and BC are 672 cm and 824 cm respectively, each answer given to the nearest cm.

41 The smallest angle of the triangle is of size 42°, to the nearest degree.

42 The parallelogram has diagonals of length 5.1 cm and 9.7 cm, correct to one decimal place.

43 The parallelogram has sides of length 6.8 cm and 10.8 cm, correct to one decimal place.

44 a When AC is 2.6 metres $\angle CAB = 20°$, to the nearest degree.

   b When AC is 2.1 metres $\angle CAB = 28°$, to the nearest degree.

45 a 479 cm   **b** 239 cm

   c 111 cm   **d** 222 cm

46 a At 5 o'clock the distance between the tip of the hour hand and the tip of the minute hand is 155 mm, to the nearest mm.

   b At 10 minutes past 5 the distance between the tip of the hour hand and the tip of the minute hand is 119 mm, to the nearest mm.

47 a The ship is 1.77 km from the lighthouse, correct to 2 decimal places.

   b The ship is 1.17 km from the coastal observation position, correct to 2 decimal places.

48 a Q is 18.5 km from P, correct to one decimal place.

   b Q is 21.0 km from the lighthouse, correct to one decimal place.

49 Ship B is approximately 15.9 km from ship A, on a bearing 164°.

50 Twelve of the steel frameworks would require a total of 260 metres of steel (to the next 10 metres).

51 The block has an area of 5270 m^2 and a perimeter of 298 metres, both answers given to the nearest integer.

52 The triangular piece that has been removed has an area of 752 mm^2 and a perimeter of 128 mm, both answers given to the nearest whole number.

53 The block has an area of 6399 m^2, to the nearest square metre.

Miscellaneous exercise eleven   PAGE 213

1 $y = 2x - 1$, B and E.

2 a $x = 17$ **b** $x = 7$ **c** $x = 2$

   d $x = -2$ **e** $x = 35$ **f** $x = 20$

3 The estimated mean = 13.9. The median lies in the $11 \rightarrow 15$ interval.

4 The mean for the whole class of students is 23.8.

5 The number I first thought of was 17.

6 Four and a half years.

7 The flagpole is of height 13.7 metres, correct to one decimal place.

8 $a = 6, b = 5, c = 19, d = 7, e = 3$.
   5, 7, 11, 13, 15, 21, 21, 25, 30, 31.

10 The boats are 7.4 km apart ninety minutes after leaving the harbour, to the nearest 100 m.

11 a 4 m   **b** 3.46 m
   c 1 m   **d** 1 m

12 The direct journey from A to C is 50.6 m shorter than the journey via B (correct to one dp).

13 Possibility 1 is of length 176 m, possibility 2 is of length 156 m and possibility 3 is of length 153 m (all to the nearest metre).

14 A mean number of nurses absent per day is 10.2, standard deviation 6.1.
   They decide to have 14 in pool.

15 a The bottom bar covers fewer years (0 – 4) than the bar above it (5 – 14).

   b Country A has more males than females for the age ranges 0 – 4, 5 – 14 and 35 – 44.

   c Country B has more females than males for the age ranges 0 – 4, 15 – 24, 25 – 34, 35 – 44, and 45 – 54.

   d Country B has approximately 4 800 000 people aged 55 and over.

   e Discuss answer and reasons with your teacher.

16 For x from 0 to 2: $y = 0.2x$.
   For x from 2 to 3: $y = 0.1x + 0.2$.
   For x from 3 to 4: $y = 0.5$.

Exercise 12A   PAGE 226

1 $x = 1, y = 6$ **2** $x = 3, y = -1$

3 $x = 4, y = 0$ **4** $x = 5, y = -2$

5 $x = -3, y = 2$ **6** $x = 7, y = 5$

7 $x = 3, y = 4$ **8** $x = 3, y = -2$

9 $x = -7, y = 10$ **10** $x = 9, y = 11$

11 $x = 5, y = -2$ **12** $x = 4, y = 7$

13 $x = 4, y = 1$ **14** $x = -3, y = 6$

15 $x = 4, y = 1$ **16** $x = 4, y = -1$

17 $x = 4, y = 3$ **18** $x = 3, y = 5$

19 $x = 7, y = 3$ **20** $A = 5, B = -2$

21 $p = 5.5, q = 3$ **22** $x = 8, y = 15$

23 $x = 10, y = 10$

24 a Equation 3: $x + y = 600$

   b Equation 4: $x - y = 140$

   c The baker baked 370 white loaves and 230 wholemeal loaves that day.

25 a Equation 1: $2x + 4y = 1758$
   b Equation 5: $x - 5y = 403$
   c Adding the number of people at the show to the number of dogs at the show gives total of 811.
26 a Equation 2: $y - x = 5$
   b Equation 4: $2x + 3y = 70$
   c The two numbers are 11 and 16.
27 a Equation 1: $x + y = 46$
   b Equation 5: $x + 0.5y = 32$
   c The piggy bank contains 18 coins that are \$1 coins and 28 that are 50 cent coins.
28 a Equation 3: $x + y = 23$
   b Equation 6: $28x + 35y = 700$
   c The seamstress bought 15 metres of material A and 8 metres of material B.
29 a Equation 3: $x + y = 25\,000$
   b Equation 6: $0.96x + 1.12y = 25\,120$
   c The investor put \$18\,000 into company X and \$7000 into company Y.
30 a Equation 1: $x + y = 35$
   b Equation 5: $3y - 2x = 15$
   c $x = 18, y = 17.$ The area of the rectangle is 306 cm^2.
31 a Equation 3: $16x + 7y = 256$
   b Equation 6: $20x + 11y = 338$
   c For five adults and three children the cost would be \$86·50.
32 The equations are $x + y = 41$ and $3y + 2x = 106$.
   The two numbers are 17 and 24.
33 The two numbers are 13 and 24.
34 The chemist should use 80 mL from bottle A and 20 mL from bottle B.
35 a $x + y = 12\,000, 1.12x + 1.05y = 13\,195$
   b $x = 8500, y = 3500$
36 450 tickets were sold for \$12 each and 1050 were sold for \$8 each.
37 The company has 16 of the 56 seaters and 9 of the 35 seaters.
38 The two numbers are 7 and 16.
39 They sold 46 jars of jam and 32 jars of relish.
40 \$75\,000 was borrowed at 14% and \$45\,000 was borrowed at 17%.
41 David answered 19 correctly.
   At least 18 questions must be correctly answered for a mark of at least 50.

Miscellaneous exercise twelve   PAGE 231

1 a • $74 - 70 \cos x$ does not equal $4 \cos x$. Rule of order not being followed.
   • If $16 = 4 \cos x$, $\cos x = 4$ not 0.25.
   • Inappropriate degree of accuracy in final answer for accuracy of given data.
   b • Not correct to use $A = 0.5ab \sin C$ for given information as the angle of known size is not <u>between</u> the two sides of known length.
   • $\dfrac{76 \times 72 \times \sin 64°}{2}$ does not equal $38 \times 36 \times \sin 32°$. Could write it as $38 \times 72 \times \sin 64°$ or as $76 \times 36 \times \sin 64°$ but not what is written in given 'solution'.
   • Final answer claims to be an area but has units of length.
   c • Given working involves the sine rule not the cosine rule as claimed.
   • Rearrangement should give $\sin \angle BCA = \dfrac{6.8 \sin 65°}{7.1}$, not what is claimed.
   • Obtuse angle should not be claimed as being a solution because $\angle BCA$ is not opposite longest side and so cannot be obtuse.
   d • Should be taking away $2 \times 5 \times 7 \cos 130°$ not adding it on.
   • $74 + 70 \cos 130°$ does not equal $144 \cos 130°$. Rule of order not being followed.
   • $144 \cos 130°$ gives a negative value for x^2, not a positive value. A negative value would then mean that x could not be determined.
   • Given value for x includes units but x is a number not a length.

2 a Nominal categorical
   b Continuous numerical
   c Ordinal categorical
   d Discrete numerical
   e Continuous numerical
   f Continuous numerical
   g Discrete numerical
   h Nominal categorical
   i Nominal categorical
   j Discrete numerical
3 a $x = 6, y = 5$ **b** $x = 3, y = 2$ **c** $x = 2, y = 3$
4 a $x = 3.6$ **b** $x = 7$ **c** $x = 15$
   d $x = 11, y = 3$ **e** $x = 3.5$ **f** $x = 14$

5 A: $y = -x + 60$ B: $y = 60$
C: $y = 2x - 60$ D: $x = 60$
E: $y = -2x + 30$ F: $y = 0.5x + 30$

6 The direct journey from A to C is 647 m shorter than the journey via B, to the nearest metre.

7 The graphs of all members of the family will have a gradient of 3.

8 The graphs of all members of the family will pass through the point (0, –7).

9 The graphs of all members of the family will have a gradient of –0.5.

10 The fire is approximately 19.7 km from lookout No.1 and 18.8 km from lookout No.2.

11 0.285 km, i.e. approximately 300 metres.

12 a The steeper the line the greater the speed. Hence we can see from the graph that the third stage was the one with the greater average speed.

   b 1st stage: 20 km/h. 2nd stage: 15 km/h. 3rd stage: 25 km/h.

   c From town A to town B the delivery van averaged 60 km/h.

   d From town B to town A the delivery van averaged 80 km/h.

   e The delivery van would need to average more than 120 km/h to arrive back at A before the cyclist got there!

13 $34 000 secure, $16 000 risky

14 $15 600 secure, $40 000 risky

15 Answers not given here. Check that your part **b** and **c** answers are the same.

16 a i The statement is incorrect. There are the same number of results to the right of the median as there are to the left of the median.

     ii The box plots do not tell us how many students were involved. Hence we cannot conclude that there were more 14 year olds than 12 year olds.

     iii The word 'much' is open to interpretation. Better to quantify rather than use words like 'much'. The reader can then decide if they wish to interpret the difference as 'much'. Instead could say, for example: The interquartile range for the 14 year olds was 7 seconds which exceeded that of the 12 year olds which was 6 seconds.

   b Discuss the reasonableness of your statements with others in your class.

Exercise 13A   PAGE 238

1 a 1   **b** 1.7   **c** –2
  d 0.5   **e** –0.75

2 Test A: 2.5, Test B: –1, Test C: 1.25, Test D: 0.2

3 Computing (1.216), Chemistry (0.278), Mathematics (–0.385), Electronics (–0.616)

4 English, Mathematics, Science, Social Studies.

5 Jill: 'Well I got 1.'
   Jill: 'The mean was zero.'
   Jill: 'Oh he got –0.25.'

Exercise 13C   PAGE 247

1 0.2266   **2** 0.6377   **3** 54.56

4 a 0.5828   **b** –0.6433   **c** 1.2265
  d –0.7388

5 a 19.5   **b** 21.9   **c** 18.7
  d 23.1

6 a 0.68   **b** 0.95   **c** 0.997
  d 0.95   **e** 0.997   **f** 0.34
  g 0.84   **h** 0.16   **i** 0.84
  j 0.16

7 a 99.7%   **b** 16%   **c** 13.5%

8 a 16%   **b** 2.5%

9 a 0.3085   **b** 0.0062

10 a 0.5   **b** 0.34

11 a 0.3085   **b** 0.2902   **c** 0.0228

12 a approx 11   **b** approx 11   **c** approx 39

13 0.0548

14 a 415   **b** 217

15 To nearest 0.5 cm: 158.5 cm, 191.5 cm

16 A/B: 78, B/C: 68, C/D: 55, D/F: 47

17 a 0.1587   **b** 7.38 am   **c** 7.33 am

18 a 2 yrs   **b** 7 yrs
  c 91 yrs   0.783

19 a approx 40   **b** 0.236

Miscellaneous exercise thirteen   PAGE 251

1 a Equation 3: $4 = 3m + c$
  b Equation 5: $19 = 8m + c$
  c The straight line equation is $y = 3x - 5$.

2 mean 24.08, median 24, mode 20, range 29.

3 Four hot dogs and six burgers would cost a total of $32.20.

4 $x = 16, y = 5$. The rectangle has an area of 80 cm².

5 Each mixed bag should contain 8 chocolates and 12 lollies.

6 a The cost of constructing 25 km of similar highway would be 155 million dollars.

b The cost of constructing 52 km of similar highway would be 317 million dollars.

7 a 167, 168, 169   **b** 166, 168, 170

8 337°, 3.4 km

9 The topmost point is 35 metres above ground (to the nearest metre).

10 a Revenue is likely to be zero for zero units sold. Thus line I is likely to be revenue line.

Costs are likely to have some fixed costs for 'set up' then an amount per unit made. Thus line II is likely to be cost line.

b Break even when $x = 40$ (then Cost = Revenue = $3200).

c I: $R = 80x$ II: $C = 2000 + 30x$

11 The yacht travelled 918 metres, to the nearest metre

12 If the agent attempts to calculate a mean he will have to decide what values to assume for the one property '$400 000 or less' and the six that are 'over $1 000 000'. Hence determining the mean presents a problem. Using the modal class will not be an indicator of *central* tendency. The median is probably the most suitable to work with. The median will be the 15th sale price and will lie in the $700 001 → $800 000

interval. The 15th will be the 'last' value in this class so he could suggest a value in this interval but towards the $800 000 end. Hence a reasonable average for the given data could be about $780 000.

In reality a lot may depend on what he wants the average for. If a customer is asking it may be more informative to show the customer the whole table as this shows the quite considerable range of prices. Just knowing an average may not be that helpful. If the average is really needed perhaps the agent should attempt to find out the actual sale prices of the 29 properties. Just how recent the sales were made could also be an important consideration.

13 a The total length of aluminium is 775 cm, to the nearest centimetre.

b The area of glass is 19 580 cm², to the nearest 10 square centimetres.

14 The distance between the hands is 471 mm, to the nearest millimetre.

15

Initials	PA	CB	JB	CC	JD	KD	LF	LJ	MJ	EK	IM	PN
Grade	C	D	A	D	B	C	F	C	B	C	B	C
Initials	RN	PP	AR	TR	VR	AS	PS	TS	BV	PV	IW	RZ
Grade	D	B	D	C	B	C	C	C	D	B	F	C

16 a 41 m (nearest metre)   **b** 26 m (nearest metre)
c 49 m (nearest metre)   **d** 328° (nearest degree)

17

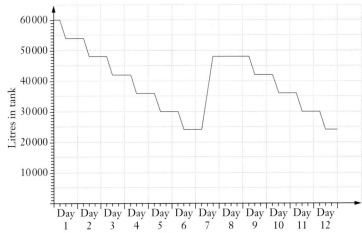

Graph of amount of water in tank over 12 day period.

18 a 144   **b** 1280
c 167   **d** 122.7, 131.3
19 a ~32.5 m   **b** ~61 m

ISBN 9780170390262

INDEX

adjacent (right-angled triangle) 171
algebraic expressions x
angle of depression 183, 184
angle of elevation 183, 184
area of a triangle 193
   given two sides and the included
      angle 195
   Heron's rule 198, 213, 234
   that are not right angled 194

bar graphs viii, 11
bearings 183–4, 185
bell shaped curve 240
bimodal ix
bivariate data 4
box and whisker diagrams (boxplots) 63–6
   versus histograms 67

categorical data 5
   displaying 6
categorical variables 5
collinear points 185
column graphs viii
compass bearings 184
continuous data 8
continuous variables 8
coordinates vii
cosine (cos) ratio 170–2
cosine rule 208–9

data, types of 4
data analysis viii, xii, 21–80
   grouped data 31
data display viii, xii, 6, 9–16, 63–8

describing distributions 36–8, 67–8, 70–3
discrete data 8
discrete variables 8, 9
distributions
   describing 36–8, 67–8, 70–3
   location, spread and shape 67–71
   skewness 68–9
dot frequency graphs viii, 11

elimination method, simultaneous
   equations 221, 224
equation of a straight line 132–3, 142–3,
   145
   given gradient and one point on the
      line 143
   given the gradient and vertical intercept
      143
   given two points that lie on the line 144
equations
   from ratios 114
   from simple interest formula 113
   solving 85–95
   to solve problems 101–120
   see also linear equations; simultaneous
      linear equations
expanding brackets x

five-number summary 63
formulae x, 85, 113
frequency histograms viii, 9–16
frequency polygons 10
frequency tables 53

Gaussian distribution 240
gradient of a straight line 127, 128–31
grouped data 31–5
   standard deviation 54–6

Heron's rule 198, 212, 234
hexapatterns 105
histograms viii, 9–16
   versus boxplots 67
horizontal lines 135
hypotenuse 171

interquartile range 63

linear equations
   simultaneous 219–30
   solving 85–95
linear relationships 125–45
   in practical situations 146–51
   see also straight line graphs
lines parallel to the x-axis 135
lines parallel to the y-axis 135
lower quartile 63

mean viii, 22–5
   notation viii, 23, 243
mean deviation 44
measures of central tendency 22, 43, 57
   see also mean; median
measures of dispersion (spread) 43–57
   see also range; standard deviation
measures of location 22
median ix, 22–3, 63
median class 32
median group 32
midpoint of each interval 32
modal class 32
modal group 32
mode ix, 22–3

negatively skewed distributions 68, 69
nominal categorical variables 5
non right-angled triangles 193–213
normal distribution 240–50
    notation for mean and standard
      deviation 245
    and percentiles 247
    and quantiles 247
    and standard deviations 240
    and z scores 242
number, use of vii
number puzzles 105–9
numerical data 8
    displaying 9–11
numerical variables 8

opposite (right-angled triangle) 171
ordinal categorical variables 5
outliers ix, 53, 65

percentage frequency 10
percentiles 247–8
pictograms viii
piecewise defined relationships 157–62
positively skewed distributions 69
primary data 4
probability xi, 240
proportional bar graphs viii
pyramids 101–4
Pythagorean theorem xi, 176

quantiles 247–8
quartiles 63

range ix, 23, 43
ratios viii
    equations from 114–15
relative frequency 10
right-angled triangles, trigonometry
    167–74

secondary data 4
similar triangles xi, 115–16
simple interest formula 113
simultaneous linear equations 219–30
    solving graphically 222
    solving using calculators 220
    solving using elimination method 221, 224
    solving using substitution method 221, 223
    word problems 225–30
sine rule 200–3, 207–8
sine (sin) ratio 170–1, 173–4
skewness of distributions 68–9
SOHCAHTOA 171
solving equations 85–95
    with brackets or fractions 88–91
    linear equations 92–5
    simultaneous linear equations 219–30
standard deviation 44, 45–52
    grouped data 54–6
    and normal distribution 238–40
    notation 47, 48, 245
    and outliers 53
    and standard scores 238–40, 242
standard normal distribution 240
standard scores 238–9, 242
statistical functions on a calculator 30–1,
    47–8
statistical investigation process 79–80
statistical tables 242
stem and leaf plots viii, 23
step graphs 158
straight line graphs 127–51
    gradient 127, 128–31
    on graphics calculators 136
    lines parallel to the axes 135
    rule 132–3
    table - rule - graph 133–4
    see also equation of a straight line
substitution method, simultaneous linear
    equations 221, 223
subtending, definition 185
summarising data viii–ix, xii, 21–9
    grouped data 31–5
summary statistics 22

tangent (tan) ratio 169–74
three figure bearings 184
triangles
    area 193–200
    cosine rule 204–7, 208–9
    sine rule 200–3, 207–8

trigonometric ratios 169–74
trigonometry for right-angled triangles
    167–87
    accuracy in trigonometric questions 183
    angles of elevation and depression
      184–5, 186–7
    applications 176–82
    bearings 183–4, 185
    calculator usage 174–5
trigonometry for non right triangles
    193–213
    area of a triangle 193–200
    cosine rule 204–7, 208–9
    sine rule 200–3, 207–8

univariate data 4
upper quartile 63

variables 4
    categorical 5
    numerical 8
variance 44, 45, 245
vertical intercept 133
vertical lines 135

$y = mx + c$ 132–3, 142–3
y-intercept 133, 134

z score 242

ISBN 9780170390262